Reviews and critical articles covering the entire field of normal anatomy (cytology, histology, cyto- and histochemistry, electron microscopy, macroscopy, experimental morphology and embryology and comparative anatomy) are published in Advances in Anatomy, Embryology and Cell Biology. Papers dealing with anthropology and clinical morphology that aim to encourage cooperation between anatomy and related disciplines will also be accepted. Papers are normally commissioned. Original papers and communications may be submitted and will be considered for publication provided they meet the requirements of a review article and thus fit into the scope of "Advances". English language is preferred.
It is a fundamental condition that submitted manuscripts have not been and will not simultaneously be submitted or published elsewhere. With the acceptance of a manuscript for publication, the publisher acquires full and exclusive copyright for all languages and countries.
Twenty-five copies of each paper are supplied free of charge.

Manuscripts should be addressed to

Prof. Dr. F. **BECK**, Howard Florey Institute, University of Melbourne, Parkville, 3000 Melbourne, Victoria, Australia
e-mail: fb22@le.ac.uk

Prof. Dr. F. **CLASCÁ**, Department of Anatomy, Histology and Neurobiology,
Universidad Autónoma de Madrid, Ave. Arzobispo Morcillo s/n, 28029 Madrid, Spain
e-mail: francisco.clasca@uam.es

Prof. Dr. M. **FROTSCHER**, Institut für Anatomie und Zellbiologie, Abteilung für Neuroanatomie,
Albert-Ludwigs-Universität Freiburg, Albertstr. 17, 79001 Freiburg, Germany
e-mail: michael.frotscher@anat.uni-freiburg.de

Prof. Dr. D.E. **HAINES**, Ph.D., Department of Anatomy, The University of Mississippi Med. Ctr.,
2500 North State Street, Jackson, MS 39216-4505, USA
e-mail: dhaines@anatomy.umsmed.edu

Prof. Dr. N. **HIROKAWA**, Department of Cell Biology and Anatomy, University of Tokyo,
Hongo 7-3-1, 113-0033 Tokyo, Japan
e-mail: hirokawa@m.u-tokyo.ac.jp

Dr. Z. **KMIEC**, Department of Histology and Immunology, Medical University of Gdansk,
Debinki 1, 80-211 Gdansk, Poland
e-mail: zkmiec@amg.gda.pl

Prof. Dr. H.-W. **KORF**, Zentrum der Morphologie, Universität Frankfurt,
Theodor-Stern Kai 7, 60595 Frankfurt/Main, Germany
e-mail: korf@em.uni-frankfurt.de

Prof. Dr. E. **MARANI**, Department Biomedical Signal and Systems, University Twente,
P.O. Box 217, 7500 AE Enschede, The Netherlands
e-mail: e.marani@utwente.nl

Prof. Dr. R. **PUTZ**, Anatomische Anstalt der Universität München,
Lehrstuhl Anatomie I, Pettenkoferstr. 11, 80336 München, Germany
e-mail: reinhard.putz@med.uni-muenchen.de

Prof. Dr. Dr. h.c. Y. **SANO**, Department of Anatomy, Kyoto Prefectural University of Medicine,
Kawaramachi-Hirokoji, 602 Kyoto, Japan

Prof. Dr. Dr. h.c. T.H. **SCHIEBLER**, Anatomisches Institut der Universität,
Koellikerstraße 6, 97070 Würzburg, Germany

Prof. Dr. J.-P. **TIMMERMANS**, Department of Veterinary Sciences, Univers
Groenenborgerlaan 171, 2020 Antwerpen, Belgium
e-mail: jean-pierre.timmermans@ua.ac.be

195
Advances in Anatomy Embryology and Cell Biology

Editors

F.F. Beck, Melbourne · F. Clascá, Madrid
M. Frotscher, Freiburg · D.E. Haines, Jackson
N. Hirokawa, Tokyo · Z. Kmiec, Gdansk
H.-W. Korf, Frankfurt · E. Marani, Enschede
R. Putz, München · Y. Sano, Kyoto
T.H. Schiebler, Würzburg
J.-P. Timmermans, Antwerpen

C. Schmidt, I. McGonnell, S. Allen
and K. Patel

The Role of Wnt Signalling in the Development of Somites and Neural Crest

With 15 Figures

Corina Schmidt

Anatomische Anstalt der LMU München
Pettenkoferstr. 11
80336 München
Germany

e-mail: corina.schmidt@med.uni-muenchen.de

Imelda McGonnell
Steve Allen

Royal Veterinary College, London

Ketan Patel

University of Reading, UK

ISSN 0301-5556
ISBN 978-3-540-77726-7 e-ISBN 978-3-540-77727-4

Library of Congress Control Number: 2008923168

© 2008 Springer-Verlag Berlin Heidelberg

This work is subject to copyright. All rights are reserved, whether the whold or part of the material is concerned, specifically the rights of translation, reprinting, reuse of illustrations, recitation, broadcasting reporduction on microfilm or in any other way, and storage in data banks. Duplication of this publication or parts thereof is permitted only under the provisions of the German Copyright Law of September 9, 1965, in its current version, and permission for use must always be obtained from Springer-Verlag. Violations are liable to prosecution under the German Copyright Law.

The use of general descriptive names, registered names, trademarks, etc. in this publication does not imply, even in the absence of a specific statement, that such names are exempt form the relevant protecttive laws and regulations and therefore free for general use.

Product liability: The publishers cannot guarantee the accuracy of any information about dosage and application contained in this book. In every individual case the user must check such information by consulting the relevant literature.

Printed on acid-free paper

9 8 7 6 5 4 3 2 1

springer.com

This work is dedicated to my beloved husband Gunter Meier

Keywords: Wnts, somite formation, neural crest induction

List of Contents

Abstract

The Wnt family of secreted signalling molecules controls a wide range of developmental processes in all metazoans. In this investigation we concentrate on the role that members of this family play during the development of (1) the somites and (2) the neural crest. (3) We also isolate a novel component of the Wnt signalling pathway called Naked cuticle and investigate the role that this protein may play in both of the previously mentioned developmental processes. (1) In higher vertebrates the paraxial mesoderm undergoes a mesenchymal-to-epithelial transformation to form segmentally organised structures called somites. Experiments have shown that signals originating from the ectoderm overlying the somites or from midline structures are required for the formation of the somites, but their identity has yet to be determined. Wnt6 is a good candidate as a somite epithelialisation factor from the ectoderm since it is expressed in this tissue. In this study we show that injection of Wnt6-producing cells beneath the ectoderm at the level of the segmental plate or lateral to the segmental plate leads to the formation of numerous small epithelial somites. We show that Wnts are indeed responsible for the epithelialisation of somites by applying Wnt antagonists which result in the segmental plate being unable to form somites. These results show that Wnt6, the only member of this family to be localised to the chick paraxial ectoderm, is able to regulate the development of epithelial somites and that cellular organisation is pivotal in the execution of the differentiation programmes. (2) The neural crest is a population of multipotent progenitor cells that arise from the neural ectoderm in all vertebrate embryos and form a multitude of derivatives including the peripheral sensory neurons, the enteric nervous system, Schwann cells, pigment cells and parts of the craniofacial skeleton. The induction of the neural crest relies on an ectodermally derived signal, but the identity of the molecule performing this role in amniotes is not known. Here we show that Wnt6, a protein expressed in the ectoderm, induces neural crest production. (3) The intracellular response to Wnt signalling depends on the choice of signalling cascade activated in the responding cell. Cells can activate either the canonical pathway that modulates gene expression to control cellular differentiation and proliferation, or the non-canonical pathway that controls cell polarity and movement (Pandur et al. 2002b). Recent work has identified the protein Naked cuticle as an intracellular switch promoting the non-canonical pathway at the expense of the canonical pathway. We have cloned chick Naked cuticle-1 (cNkd1) and demonstrate that it is expressed in a dynamic manner during early embryogenesis. We show that it is expressed in the somites and in particular regions where cells are undergoing movement. Lastly our study shows that the expression of *cNkd1* is regulated by *Wnt* expression originating from the neural tube. This study provides evidence that non-canonical Wnt signalling plays a part in somite development.

1 Introduction

1.1 Wnts and Development

The family of Wnt genes consists of at least 19 members in vertebrates. *Wnts* have multiple roles during normal development and aetiology of diseases (Wodarz and Nusse 1998; Moon et al. 2002, 2004). These genes encode for glycoproteins that are released in the intercellular space, acting as intercellular mediators (Moon et al. 1997). By binding to their special receptors, the Frizzled (Fz) receptors, they are able to activate at least three different pathways: the canonical, the non-canonical and the Ca2+ pathway (Fig. 1). The Wnt family has been subdivided, using functional assays, into at least two subclasses. In *Xenopus*, Wnt1, Wnt3a and Wnt8 are able to induce axis duplication in embryos, whereas Wnt4, Wnt5a and Wnt11 cannot (Du et al. 1995). Those Wnts that are able to induce axis duplication are thought to act canonically; those who cannot are thought to act non-canonically. In the literature, therefore, some Wnts are called canonical Wnts while others are called non-canonical Wnts, suggesting that a special Wnt always activates the same signalling pathway. It is unlikely that the specificity dictating cellular responses resides solely in the Wnt ligands because no one has reported sequence or structural motifs in Wnts that predict their activities in these cellular and embryonic assays. Further evidence showing that cellular responses cannot be predicted based solely on the identity of the Wnt comes from the finding that the "non-canonical" Wnt5 is able to induce axis duplication and stabilisation of β-catenin if some Fz receptors are also present; in other words, it is able to act canonically (He et al. 1997; Kühl et al. 2000). We should therefore act on the assumption that any Wnt is able to activate different signalling pathways depending on the expression of Fz receptors and on the presence of different signal-interfering molecules in the cellular environment.

Wnts are involved in numerous developmental processes. They participate in the primary axis formation during gastrulation (Takada et al. 1994; Yoshikawa et al. 1997; Sokol 1999; Yamaguchi et al. 1999), in the development of the extremities (McMahon et al. 1992; Parr et al. 1993, 1998; Parr and McMahon 1995), in the development of the kidney (Kispert et al. 1996, 1998; Lescher et al. 1998), in the mesodermal patterning, somitogenesis, cardiogenesis and brain morphogenesis (Wilkinson et al. 1987; Roelink et al. 1990; Thomas and Capecchi 1990; Parr et al. 1993; Pandur et al. 2002a; Aulehla et al. 2003; Schmidt et al. 2004; Linker et al. 2005). Additionally, they play an important role in the compartmentalisation of the somites (Marcelle et al. 1997; Fan et al. 1997; Ikeya and Takada 1998; Borycki et al. 2000; Wagner et al. 2000), and in the subsequent myogenesis (Münsterberg et al. 1995; Tajbakhsh et al. 1998; Ridgeway et al. 2000; Wagner et al. 2000; Schmidt et al. 2000; Geetha-Loganathan et al. 2005, 2006). In adults, misregulation of the Wnt signalling pathway has been linked to various human cancers, including colon and hepatocellular carcinomas, leukaemia and melanoma (Moon et al. 2004; Gregorieff

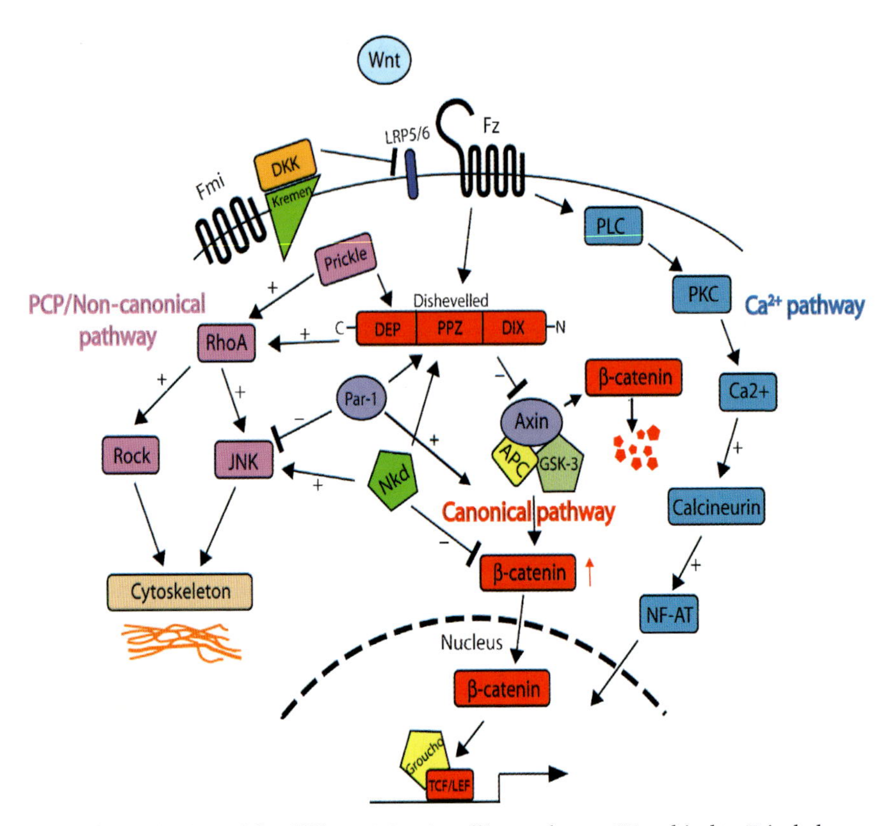

Fig. 1 Schematic view of the different Wnt signalling pathways. Wnts bind to Frizzled receptors. LRP5 and LRP6 are members of the low-density lipoprotein receptor-related family and act as essential co-receptors of Wnt signalling. Dickkopf (Dkk) antagonises Wnt action by blocking the LRP co-receptor and induction of LRP endocytosis in co-operation with Kremen proteins. The non-canonical pathway leads to the induction of RhoA/Rho kinase (ROCK) and c-Jun kinase (JNK), both interacting with cytoskeletal components of the cell, thereby influencing for example, cell migration or cell division. In the canonical pathway a multiprotein complex containing Axin, APC and GSK-3 is inhibited via Dishevelled. This leads to an accumulation of β-catenin in the cell. β-Catenin is able to enter the nucleus where it replaces Groucho and complexes with TCF/LEF transcription factors, turning them into transcriptional activators thereby inducing gene expression. The Wnt/Ca^{2+} pathway leads to release of intracellular Ca^{2+} via activation of phospholipase C (PLC) and protein kinase C (PKC). Elevated Ca^{2+} is able to activate the phosphatase calcineurin, which leads to dephosphorylation of NF-AT. NF-AT is able to enter the nucleus and interact with gene expression

and Clevers 2005; Radtke and Clevers 2005; Reya and Clevers 2005). Most colorectal cancers contain sporadic truncations of the adenomatous polyposis coli (APC) tumour suppressor. The most frequent of the cancer-causing mutations yield truncated APC proteins that are unable to bind Axin or degrade β-catenin (Kinzler and Vogelstein 1996), resulting in the sustained expression of Wnt target genes. Wild-type, but not mutant, APC also interacts with the microtubule cytoskeleton and can help maintain chromosomal stability through association with the kinetochore of metaphase chromosomes (Bienz and Hamada 2004). A fraction of tumours with wild-type APC contain oncogenic stabilising mutations that prevent phosphorylation of β-catenin, thus allowing it to accumulate in the nucleus independently of a Wnt signal. Remarkably, some tumours may arise as a consequence of selective loss of Wnt-/β-catenin signalling. One-third of human sebaceous gland tumours therefore contain a dual-site mutation in LEF-1 that impairs its ability to bind β-catenin and causes it to function as a dominant-negative inhibitor of Wnt-signalling (Takeda et al. 2006).

Extracellularly located Wnt proteins activate—as mentioned above—at least three different intracellular pathways by binding the Fz seven transmembrane-spanning receptors (Bhanot et al. 1996; He et al. 1997; Borello et al. 1999; Cauthen et al. 2001). These pathways are (1) the β-catenin pathway (canonical Wnt pathway), which activates target genes in the nucleus; (2) the planar cell polarity (PCP) pathway (PCP/JNK pathway or non-canonical Wnt pathway) and (3) the Wnt/Ca2+ pathway (Fig. 1). In addition to Wnt/Fz interactions, Wnt signalling also requires the presence of a single-pass transmembrane molecule, a member of the low-density lipoprotein receptor (LDLR)-related protein (LRP) family, identified as the gene *Arrow* in *Drosophila* (Wehrli et al. 2000) and as *LRP5* or *LRP6* in vertebrates (Tamai et al. 2000). The transport of LRP to the cell surface requires a specific molecule called Boca in *Drosophila* and Mesd in mice (Culi and Mann 2003; Hsieh et al. 2003). It has been proposed that Wnt molecules bind to LRP and Fz to form a receptor trimeric complex (Tamai et al. 2000) which is able to transduce the Wnt signal to the cytoplasm of the cell. Furthermore, the importance of LRP is underscored by the finding that extracellular inhibitors of Wnt signalling such as Wise (Itasaki et al. 2003) and Dickkopf (Dkk) (Glinka et al. 1998) bind to LRP. Dkk proteins, the best characterised of the secreted Wnt signalling inhibitors, have not been found in invertebrates, but mice and humans have multiple Dkk genes (Krupnik et al. 1999; Monaghan et al. 1999). In particular, Dickkopf-1 is a potent inhibitor of Wnt signalling (Glinka et al. 1998). It binds to LRP with high affinity (Bafico et al. 2001; Semenov et al. 2001) and to another class of transmembrane molecules, the Kremens (Mao et al. 2002). By forming a complex with LRP and Kremen, Dkk promotes the internalisation of LRP, thereby making it unavailable as a Wnt receptor. The inhibitory function of Dkks depends on the presence of appropriate Kremen proteins. For example, Dickkopf-2 requires Kremen-2 in order to inhibit Wnt signalling (Mao and Niehrs 2003). Likewise, Kremen-2 promotes the inhibitory activity of Dickkopf-4 (Mao and Niehrs 2003).

Furthermore, in the process of Wnt signal transduction the Fz receptors compete with the Fz-related proteins (Sfrps). These are secreted proteins related to the Fz receptors but without a transmembrane domain. These include Sfrp1 (also known as hFRP, FrzA and SARP2), Sfrp2 (also called SARP1 and SDF-5), Frzb (Sfrp3 and Fritz), Sfrp4, SARP3, Sizzled and Crescent, which have so far been characterised in humans, mice, chick and *Xenopus* (Shirozu et al. 1996; Finch et al. 1997; Mayr et al. 1997; Melkonyan et al. 1997; Pfeffer et al. 1997; Rattner et al. 1997; Salic et al. 1997; Hoang et al. 1998; Leimeister et al. 1998; Lescher et al. 1998; Lee et al. 2000). Cerberus, WIF-1 protein and—as already described—Dickkopf proteins are also secreted and can antagonise Wnt signalling (Glinka et al. 1998; Hsieh et al. 1999;

Piccolo et al. 1999). Together with the Sfrps these molecules form one group of proteins that may be critical in modulating Wnt function (Lin et al. 1997; Moon et al. 1997; Ladher et al. 2000; Terry et al. 2000). It is worth mentioning that, in addition to Fz and LRP5/6, there are two other Wnt co-receptors: Ror2, which is bound by Wnt5a and activates Jun N-terminal kinase (JNK) (Oishi et al. 2003; Mikels and Nusse 2006) and the tyrosine kinase Derailed/Ryk (Yoshikawa et al. 2003). Their mechanism of signal transduction is poorly understood.

The Wnt–Fz interaction leads to activation of Dishevelled (Dsh), which itself is able to activate both the canonical and the non-canonical signalling pathway (Veeman et al. 2003). The vertebrate Dsh family includes three homologues, Dsh-1, Dsh-2 and Dsh-3, which exhibit overlapping patterns of expression and highly conserved domain structure (Sussman et al. 1994; Klingensmith et al. 1996; Tsang et al. 1996; Semenov and Snyder 1997). The N-terminal DIX domain of Dsh contains a sequence also found in Axin and mediates the interaction between Dsh molecules and Axin (Kishida et al. 1999). The DIX domain is necessary for canonical signalling (Li et al. 1999). The central PDZ domain mediates the interaction of Dsh with many different proteins including Axin and Naked cuticle (Nkd) (Fig. 1; reviewed in Wharton 2003). Additionally, two kinases which have been shown to phosphorylate Dsh, casein kinase Iε (CKIε) and casein kinase II (CKII), interact with Dsh through the PDZ domain (Willert et al. 1997; Peters et al. 1999; Sakanaka et al. 1999). CKIε stimulates Dsh activity in the canonical Wnt pathway, but inhibits Dsh activity in the JNK pathway (Cong et al. 2004) possibly by altering the conformation of the C-terminus of Dsh. The C-terminal DEP domain of Dsh which binds to Prickle (Tree et al. 2002) has been shown to be required for the activation of the JNK pathway (Li et al. 1999; Moriguchi et al. 1999) and for the establishment of planar cell polarity in *Drosophila* (Boutros et al. 1998). It has been shown in *Xenopus* that Prickle binds to *Xenopus* Dsh as well as to JNK. This suggests that Prickle plays a pivotal role in connecting Dsh function to JNK activation.

In the absence of Wnt stimulation, β-catenin, an intracellular signalling component, is degraded via phosphorylation-dependent ubiquitination and proteolysis. This process occurs in the Axin complex composed of the scaffolding protein Axin, tumour-suppressor protein APC, glycogen synthase kinase-3 (GSK-3) and casein kinase I (CKI). The latter two, GSK-3 and CKI, sequentially phosphorylate β-catenin (Liu et al. 2002), targeting it for subsequent degradation (Polakis 2002). In the presence of Wnts and activation of the canonical pathway, Dsh inhibits a complex that consists of GSK-3, Axin, APC protein and β-catenin. This leads to an accumulation of unphosphorylated β-catenin, the key molecule of a complex signalling network (Gumbiner 1995; Willert and Nusse 1998). After the entry into the nucleus, β-catenin binds to TCF/LEF (T cell factor/lymphocyte enhancer factor) transcription factors (Huber et al. 1996; Eastman and Grosschedl 1999). TCF provides sequence-specific binding activity and, in the absence of nuclear β-catenin, partners with the transcriptional repressor Groucho to form a repressive complex and block transcription of Wnt target genes (Cavallo et al. 1998; Chen et al. 1999). When β-catenin enters the nucleus, it directly replaces Groucho from its binding of TCF and converts the complex to a transcriptional activator, thereby effecting the transcription of Wnt target genes (Daniels and Weis 2005). The Wnt-activated TCF/LEF transcription factors regulate the expression of numerous genes that encode for a variety of proteins, including other transcription signalling proteins and cytoskeletal components (Molenaar et al. 1996; McKendry et al. 1997; Bienz 1998).

In the non-canonical (PCP) Wnt pathway, Fz activates JNK via the recruitment of small GTPases of the rho/cdc42 family and directs asymmetric cytoskeletal organisation and co-ordinated polarisation of cells within a plane of epithelial sheets. The PCP pathway branches at the level of Dsh from the canonical pathway. The choice of which of the two pathways is used depends on the receptor profile of the cell and the activated protein domain of Dsh, but it is also modulated by the presence of special molecules in the cytoplasm of the cell itself. The precise mechanism by which Wnt signalling "switches" between the canonical and non-canonical Wnt pathway is, however, unclear; apart from the above-mentioned CKIε and Prickle, proteins such as PAR-1 and Nkd seem to be involved. PAR-1 has been identified as a kinase that binds and phosphorylates Dsh in a region amino-terminal to the PDZ domain (Sun et al. 2001). Wnt signalling increases endogenous PAR-1 kinase activity, and this subsequently potentiates the canonical Wnt pathway at the level upstream of Axin and β-catenin. In contrast, activation of PAR-1 inhibits Dsh-mediated JNK activation and therefore the co-ordination of planar cell polarity (Sun et al. 2001). Thus, PAR-1 is a positive regulator of the Wnt/β-catenin pathway and an inhibitor of the JNK pathway. It has to be mentioned that recent RNA interference (RNAi) studies in *Drosophila* cells cast some doubt on this conclusion (Matsubayashi et al. 2004). Moreover, it has been suggested that PAR-1 depletion in *Xenopus* does not affect known targets of canonical Wnt signalling but rather is required for convergent extension movements associated with PCP signalling (Kusakabe and Nishida 2004). A study from Ossipova et al. (2005) resolves this contradiction. They show not only that PAR-1 is essential for both establishment of the organiser (as a canonical Wnt function) and convergent extension but also that these two functions—outputs of the different Wnt pathways—are mediated by different isoforms of the PAR-1 kinase. In the end they conclude that there are two classes of PAR-1 kinase, each of them associated with a specific Wnt pathway (canonical and PCP pathway).

Nkd is a protein which is able to inhibit the canonical Wnt pathway via binding of Dsh but also to activate the non-canonical pathway via stimulation of JNK (Yan et al. 2001; Rousset et al. 2001; Wharton et al. 2001). Nkd was the first Wnt antagonist found to be induced by the Wnt pathway (Zeng et al. 2000; Schmidt et al. 2006). The activation of a special Wnt pathway is therefore a result of a very complex interaction of different molecules at different levels of the signalling cascade. The activation of the third pathway, the Wnt/Ca^{2+} pathway, which is still controversial and may partly overlap with the PCP pathway, leads to a release of intracellular calcium. The elevated Ca^{2+} level activates the phosphatase calcineurin. Once calcineurin has been activated, it removes phosphate moieties from numerous proteins. One major target is the transcription factor NFAT. Phosphorylated NFAT is unable to translocate to the nucleus. Once the phosphate group has been removed, however, it enters the nucleus and initiates gene transcription (Huelsken and Behrens 2002).

Through the work in our lab we are gaining some insight how Wnt signalling is involved in the process of somite formation and neural crest induction. But before we go to the results of this work, a short overview is given concerning how these processes occur during normal development of the vertebrate embryo.

1.2 Somite Formation

The skeletal muscles of the body originate from the somites that bud off as segmental spheres from the unsegmented paraxial mesoderm, designated as segmental plate in avian embryos or presomitic mesoderm in mice embryos, in a cranial-to-caudal progression (for an overview see Christ and Ordahl 1995; Christ et al. 1998). Somitogenesis could be subdivided into two different processes, the segmentation and the epithelialisation of the paraxial mesoderm.

During the first phase of somitogenesis somite formation is governed by a molecular oscillator termed the segmentation clock (Cooke and Zeeman 1976; Schnell and Maini 2000). This clock controls oscillating patterns of gene expression within the segmental plate thereby producing periodicity within cells destined for somite formation. This oscillating gene expression induces the formation of clusters of mesenchymal cells within the segmental plate, the so-called somitomeres (Meier 1979; Jacobson 1988). Different genes, e.g. *Hairy* and *Lunatic fringe*, are expressed in a periodic manner. These cyclic genes are expressed as a wave sweeping the segmental plate in a caudal-to-cranial fashion, once during the formation of each somite (Palmeirim et al. 1997; Forsberg et al. 1998; Aulehla and Johnson 1999). *C-Hairy1* expression first appears in a broad domain extending caudally down to the prospective tail-bud region. This expression domain progressively sweeps anteriorly while narrowing to finally cover a half-somite size domain which gives rise to the caudal half of the forming somite where expression is then maintained. The dynamic wave-front of expression is independent of cell movements and does not result from the propagation of a signal in the plane of the segmental plate but rather corresponds to an intrinsic property of the cells there. As *c-Hairy1* expression sweeps across the length of the entire segmental plate during the formation of every somite, all cells within the segmental plate will experience a *c-hairy1* on and off phase during this interval (Palmeirim et al. 1997).

The majority of the cycling genes found in vertebrates are intimately related to the Notch signalling pathway (Freitas et al. 2005). Notch is a transmembrane receptor which is able to interact with two sets of transmembrane ligands, namely Delta and Serrate (Artavanis-Tsakonas et al. 1999). Upon ligand binding, Notch undergoes proteolytic cleavage, leading to the translocation of its intracytoplasmic domain into the nucleus where it activates the transcription of downstream genes such as those of the Enhancer of split complex in *Drosophila* or HES genes in vertebrates. The involvement of the Notch signalling pathway in somitogenesis was first revealed with the finding that *Notch1* and its ligand *Delta-like1*, both expressed in the presomitic mesoderm of mouse embryos, play important roles in boundary formation of somites (Conlon et al. 1995; Hrabe de Angelis et al. 1997).

The importance of this pathway for somitogenesis was supported by numerous other studies showing that dysfunction of Notch signalling leads to segmentation defects (Barrantes et al. 1999; Conlon et al. 1995; Evrard et al. 1998; Hrabe de Angelis et al. 1997; Jen et al. 1997; Bessho and Kageyama 2003). Nevertheless, it must be mentioned that though segmentation was severely affected in these mutants it was never inhibited completely, suggesting that maybe other signalling molecules are involved in this process. Indeed, the Wnt pathway has also been suggested to be implicated in the segmentation clock. *Axin2* exhibits an oscillatory expression in the paraxial mesoderm of mice. The expression of *Axin2* is still present in Notch pathway mutants whereas *Axin2* and *Lunatic fringe* oscillation are disrupted in Wnt3a mutants. This indicates that the Wnt signalling pathway could act upstream of the Notch signalling pathway (Aulehla et al. 2003). The periodic signal generated by the clock is converted into the appearance of regular spaced somite boundaries through the interaction with the so-called wavefront by a mechanism involving decreasing posterior-to-anterior gradients of fibroblast growth factor 8 (FGF8) (Dubrulle et al. 2001) and Wnt3a (Aulehla et al. 2003) and an increasing posterior-to-anterior gradient of retinoic acid (Moreno and Kintner 2004).

Numerous papers extended this work by showing that retinoic acid is also involved in left–right symmetry during somite formation (Vermot et al. 2005; Vermot and Pourquie 2005; Kawakami et al. 2005; Sirbu and Duester 2006). Actually Moreno and Kintner mentioned in their work (2004) that FGF transfers its effect on paraxial cells via activation of the MAPK/ERK pathway. This has been demonstrated by Delfini and co-workers (2005) in chicken embryos. They showed that the FGF8 gradient is translated into graded activation of the extracellular signal-regulated kinase (ERK)/mitogen-activated protein kinase (MAPK) pathway in the segmental plate (Delfini et al. 2005). ERK itself regulates the motility of cells within the segmental plate thereby enabling these cells to undergo proper segmentation.

During the second phase of somitogenesis which is called epithelialisation, a mesenchymal-to-epithelial transformation of the cranial segmental plate takes place leading to the formation of epithelial somites (Fig. 2). In this process Paraxis plays an important role. Paraxis is a basic helix-loop-helix transcription factor which is expressed in the cranial segmental plate, the epithelial somites and also in the dermomyotome of the differentiated somites. We were able to show that removal of the overlying surface ectoderm or implantation of a barrier between the neural tube and paraxial mesoderm at the level of the caudal segmental plate led to a loss of *Paraxis* expression combined with a loss of the epithelialisation of the paraxial mesoderm at the operation level (Sosic et al. 1997). We suggested therefore that a factor within the ectoderm induces the expression of *Paraxis* in the cranial segmental plate enabling the epithelialisation of the paraxial mesoderm and thereby the formation of the epithelial somites. Interestingly, the Paraxis knockout mouse has no epithelial somites but shows segmentation of the paraxial mesoderm (Burgess et al. 1996). This shows that segmentation and epithelialisation are two different processes leading to proper somite formation. Palmeirim and colleagues (1998) confirmed this by showing that explants from presomitic chicken

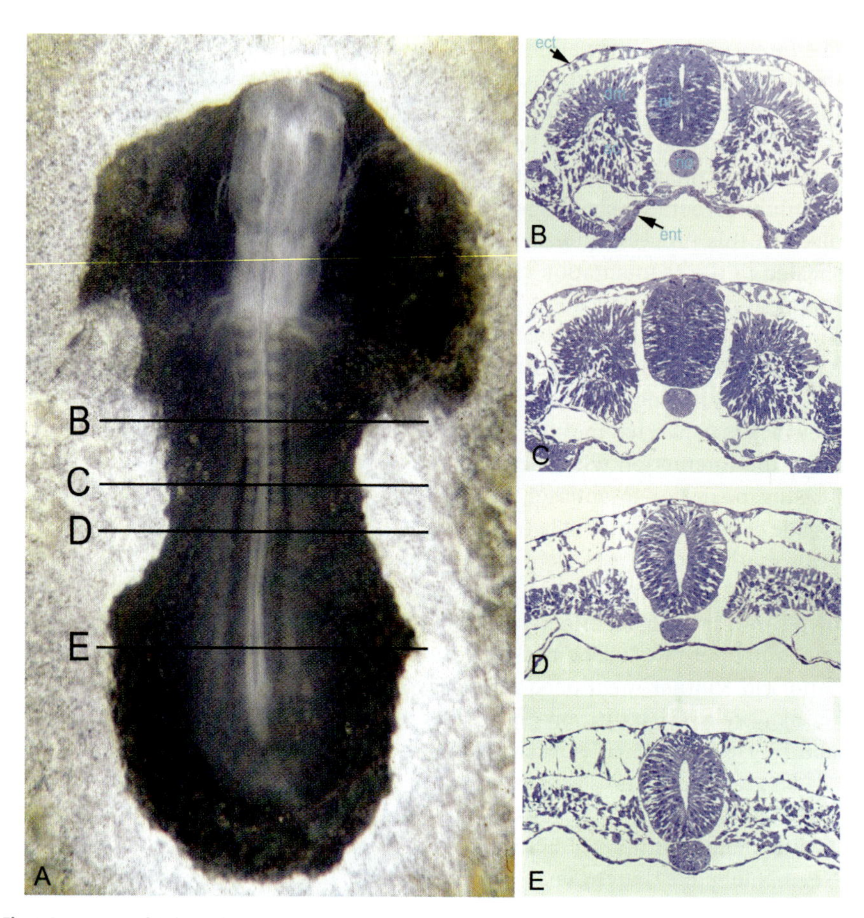

Fig. 2 A HH12 chick embryo. **B–E** Transverse sections of the embryo shown in **A**. **B** Section at the level of the differentiated somite (*ect*, ectoderm; *ent*, entoderm; *dm*, dermomyotome; *sc*, sclerotome; *no*, notochord; *nt*, neural tube). The somite is differentiated in an epithelial dermomyotome (*dm*) and a mesenchymal sclerotome (*sc*). **C** Section at the level of the epithelial somite. The epithelial somite shows a mesenchymal core (somitocoel) surrounded by epithelial somitic cells. **D** Section at the level of the cranial segmental plate. The cells of the cranial segmental plate show a partial epithelialisation in the dorsal and ventral part. **E** Section at the level of the caudal segmental plate. The cells of the caudal segmental plate are mesenchymal

mesoderm without overlying ectoderm continued to lay down stripes of *c-Delta-1* expression, although epithelialisation was blocked. These results suggested on one hand that somite rostrocaudal patterning is an autonomous property of the presomitic mesoderm; on the other hand, it showed that segmentation is not necessarily coupled with somite formation (Palmeirim et al. 1998).

The newly formed somites consist of an epithelial ball enclosing the somitocoel (Fig. 2; Mestres and Hinrichsen 1976). The somitocoel consists of mesenchymal cells which take part in the formation of the intervertebral discs, intervertebral joints and ribs (Huang et al. 1994, 1996, 2000). The ventro-medial part of the somite undergoes an epithelio-mesenchymal transition and forms the sclerotome (Fig. 2), which—together with the cells of the somitocoel—gives rise to the vertebral column and ribs (Evans 2003). Development of the sclerotome, marked by the expression of *Pax1*, is regulated by the notochord and the floor plate of the neural tube (Ebensperger et al. 1995; Balling et al. 1996). Pax1 encodes a transcription factor that is necessary for the proper formation of the vertebrate bodies and the intervertebral discs (Wallin et al. 1994). *Pax1*-deficient mice lack vertebral bodies and intervertebral discs (Wallin et al. 1994; Balling et al. 1996). The remaining epithelial, dorso-lateral part of the somite forms the dermomyotome (Fig. 2). The dermomyotome gives rise to the musculature of the body and extremities as well as to the dermis of the back. Under the influence of the neural tube and the overlying ectoderm, *Pax3* is expressed in this dorsal somitic compartment (Dietrich et al. 1997). In addition, *Pax3* is expressed in the neural tube and in the cells of the caudal segmental plate. Later on, expression of *Pax3* in somites is only found in the dermomyotome (Williams and Ordahl 1994). There the expression is found in the medial and lateral part, which includes the muscle progenitor cells of the back and the extremities (Christ et al. 1974, 1977; Ordahl and Le Douarin 1992).

Under the influence of neighbouring structures, the cells of the dermomyotome form the underlying myotome. In the chicken embryo the cells of the medial myotome express the myogenic determination factors *Myf-5* and *MyoD*. These cells form the epaxial myotome giving rise to the musculature of the back (Kiefer and Hauschka 2001). The lateral part of the myotome, expressing *Pax3* and *Myf-5*, is the source of the hypaxial musculature (Denetclaw et al. 1997; Denetclaw and Ordahl 2000; Huang et al. 2000).

In vivo and in vitro work has shed light on the molecules involved in somite patterning. Wnts have been recognised as the major players in the establishment of the epithelial dermomyotome dorsally and Shh and Noggin as the major ventralising signals. Shh and Noggin produced and secreted by the notochord are the forces driving the epithelial transition in the ventral somite leading to the formation of the mesenchymal sclerotome (Brand-Saberi et al. 1993; Fan and Tessier-Lavigne 1994; Dietrich et al. 1997; Borycki et al. 1998; Christ et al. 2004). Wnts produced by the neural tube and surface ectoderm are not only involved in regulating the properties of the medial and lateral halves of the dermomyotome including its proliferation but also in the formation of the underlying myotome (Münsterberg et al. 1995; Wagner et al. 2000; Olivera-Martinez et al. 2001; Galli et al. 2004). Interestingly, regarding the myotome, Wnts and Shh synergise to induce expression of myotomal markers such as *Myf5* via different regulation of *Gli* genes (Borycki et al. 2000) and activation of the Wnt/β-catenin pathway (Borello et al. 2006).

1.3 Induction, Delamination and Migration of Neural Crest

The neural crest is a cell population which has the ability to differentiate into a vast array of cell types and to migrate from its point of origin over considerable distances along defined routes to specific destinations (Le Douarin 2004). Neural crest cells originate from the ectoderm. The epithelial precursors of this tissue lie between cells that will become epidermis and neural epithelium. In most vertebrates, the neural crest arises from the entire length of the neuraxis starting caudal to the prospective diencephalon (Huang and Saint-Jeannet 2004). After the ectoderm has been patterned into the epidermal, neural and neural crest fates, the neural crest undergoes an epithelial-to-mesenchymal transition allowing it to delaminate from the neural epithelium and migrate as undifferentiated precursor cells. Migration is terminated when the cells reach their final destination at which point the cells differentiate. Neural crest cells give rise not only to components of the nervous system including sensory neurons and glia but also to non-neuronal cells such as bone, cartilage and pigment cells. A major area for research has focussed on establishing how one cell type can differentiate into so many tissues. Initially, fate mapping studies seemed to explain this conundrum. They indicated that, during normal development, neural crest fates were restricted to specific levels along the anterior–posterior axis. For example, cartilage and bone only developed from the cranial neural crest, not the trunk crest. These results implied that the neural crest had intrinsic information regarding their differentiation behaviour. McGonnell and Graham (2002), however, showed that avian trunk neural crest cells cultured in appropriate media are able to form both cartilage and bone cells and when they are implanted into the developing head, they contribute to cranial skeletal components. Thus, the heterotopic grafts of neural tissues showed that the fate of the tissue was plastic and that the differentiation programme was under the influence of environmental signals (McGonnell and Graham 2002).

Considerable efforts have been expended in the search for environmental signalling molecules that regulate differing aspects of neural crest cell behaviour, ranging from the induction of this unique tissue to its ability to undergo an epithelial-to-mesenchymal transition and its capacity to migrate and finally to differentiate. One family of secreted signalling factors that can influence most of these steps are the Wnt proteins. In the following we focus on the role of Wnt signalling in the induction, delamination and migration of the neural crest. It is also sometimes necessary to discuss the role of other signalling proteins such as BMP (bone morphogenetic protein) or FGF (fibroblast growth factor) since they are often intimately associated with the Wnts in the development of the neural crest.

1.3.1 Induction of the Neural Crest

The neural crest arises at the border of neural and non-neural ectoderm. The induction of the neural crest takes place while the cells are still epithelial, and this process can be monitored by the expression of neural crest-specific genes. In amphibians and birds one of the first steps of neural crest induction is the expression of the transcription factor *Slug* (Nieto et al. 1994; Mayor et al. 1995). The cloning and characterisation of three *Xenopus Slug* promoters revealed that one of the common regulatory regions contains a Lef/β-catenin binding site necessary for specific expression. As the Lef/β-catenin binding complex is a downstream effector of Wnt signalling, these results suggest strongly that Wnts are directly involved in neural crest cell induction (Vallin et al. 2001). *RhoB*, a member of the rho gene family of small guanosine triphosphate (GTP)-binding proteins, is expressed in early neural crest cells during delamination, which is a prerequisite for the following migration of these cells (Liu and Jessell 1998). HNK-1, a glycoprotein/glycolipid epitope, is expressed in migrating neural crest cells. Although not specific for the neural crest, it is a good marker for these cells, both as they emerge from the neural tube and proceed along their pathway of migration (Bronner-Fraser 1986).

The use of markers has enabled significant advances to be made in elucidating the mechanisms that regulate all aspects of neural crest behaviour. Tissue manipulations and molecular investigations have shown that interactions between non-neural ectoderm and naïve neural plate tissue are required to induce the formation of neural crest cells (Selleck and Bronner-Fraser 1995; Dickinson et al. 1995). The nature of the inductive signal has been deduced using transfilter studies performed on HH stage 4 (Hamburger and Hamilton 1951) neural plates of chick embryos which led to the conclusion that the signal emanating from the non-neural ectoderm inducing neural crest formation must be diffusible (Selleck and Bronner-Fraser 2000). Earlier studies in chick by Liem et al. (1995) showed that BMP-4 and BMP-7 were able to induce neural crest cell production from chick neural explants in vitro (Liem et al. 1995). In contrast, other experiments using chicken neural plate showed that BMP-4 is unable to induce neural crest (Garcia-Castro et al. 2002). Another apparent contradiction was the finding that in *Xenopus*, BMP antagonists induce ectopic expression of neural crest markers (Mayor et al. 1995). These apparently contradictory findings can be reconciled by findings from *Xenopus* animal cap assays showing that neural crest induction requires both BMP-4 and its antagonist Noggin (Baker and Bronner-Fraser 1997; Marchant et al. 1998). A combined action of both molecules is necessary to produce a morphogen gradient of BMP activity where low BMP activity induces the neural plate, intermediate levels induce the neural crest, and high BMP activity induces the epidermis. This gradient is produced by differing concentrations of Noggin.

Studies from La Bonne and Bronner-Fraser (1998) extended this idea of neural crest induction. They showed using the *Xenopus* model that in neural plate tissue Chordin (a potent BMP antagonist) alone is insufficient to account for neural crest induction. However, they found that such Chordin-induced neural plate tissue could be induced to develop the neural crest by members of the FGF and Wnt families. Furthermore, they showed that endogenous Wnts play an important role in neural crest formation. Blocking of Xwnt8 function via injection of dominant-negative Xwnt8 in the animal pole of two-cell-stage embryos led to an inhibition of the expression of the neural crest marker *Xslug*. This group therefore suggested a two-signal model where BMP signalling at the lateral edge of the neural plate provides an initial, weak specification of neural crest fate but additional signals such as Wnts or FGFs enhance and maintain this induction. They stated that Wnt signalling acts directly to induce neural crest cells, whereas neural crest induction via FGF may be mediated by Wnt signals (La Bonne and Bronner-Fraser 1998). This could explain the fact that paraxial mesoderm has been shown to be able to induce neural crest since FGFs are expressed in that tissue (Bonstein et al. 1998; Monsoro-Burq et al. 2003) and may activate *Wnts* thereby promoting neural crest induction. Consistent with the two-signal model of neural crest induction was the observation that in *Xenopus* both intermediate levels of BMPs and FGF, RA or Wnts are required for the induction of neural crest cells (Villanueva et al. 2002). Many other studies have found that the over-expression of *Wnts* induced neural crest formation (Saint-Jeannet et al. 1997; Chang and Hemmati-Brivanlou 1998) or that the inhibition of Wnt signalling blocked neural crest induction (Bastidas et al. 2004; Deardorff et al. 2001). Recent studies have shown that Wnts not only play an important role during the early stages of neural crest induction but also act later providing patterning cues during crest cell migration to specify which cells adopt pigment cell fate (Lewis et al. 2004).

In summary, we conclude that neural crest induction which occurs at the border of neural and non-neural ectoderm is induced by the concerted action of different factors and that the Wnts play a prominent role in this process.

1.3.2 Delamination and Migration of the Neural Crest

After their induction, neural crest cells delaminate from the neural tube and migrate as undifferentiated precursors to their final destinations. The delamination process involves an epithelio-mesenchymal transition (EMT) of neural crest cells.

The EMT is a complex process involving disruption of the basal lamina, dissociation of cell adhesions, changes in extracellular matrix, translocation of the cell body and directed migration (Hay 1995; Savagner 2001). In avian embryos, trunk neural crest begins to emigrate at the level of epithelial somites, and this process is fully underway at the level of compartmentalised somites. Sela-Donenfeld and Kalcheim (1999) showed that *BMP-4* expression is localised to the dorsal aspect of the neural tube along its entire length. However, there is a gradient of expression of the BMP inhibitor *Noggin*, with lowest levels in the rostral aspect and highest in the caudal region of the neural tube. This leads to a region of high BMP-4 activity (uniform levels of BMP-4 and a low Noggin concentration) at the level of the compartmentalised somites where the neural crest delaminates. A lower level of BMP-4 activity, however, is found at the level of the unsegmented paraxial mesoderm (constant levels of BMP-4 and a high Noggin concentration) where delamination does not occur. Disturbance of this BMP–Noggin gradient—by grafting Noggin-producing cells dorsal to the neural tube at levels opposite of the newly formed somites—inhibited emigration of HNK-1-positive crest cells, which instead accumulated in the neural tube. This Noggin-dependent inhibition was overcome by concomitant treatment with BMP-4. Furthermore, BMP-4, when added alone, also accelerated cell emigration compared to untreated controls. The authors therefore concluded that the co-ordinated activity of Noggin and BMP-4 in the dorsal neural tube triggers the delamination of *Slug*-expressing neural crest cells (Sela-Donenfeld and Kalcheim 1999). The same group also showed that the expression of *Noggin* in the neural tube is controlled by the paraxial mesoderm (Sela-Donenfeld and Kalcheim 2000). That BMP promotes delamination was shown indirectly by Coles et al. (2004). They demonstrated that the chick homologue of Crossveinless 2 (Cv-2), a *Drosophila* gene that was identified as a promoter of BMP-like signalling, is also involved in neural crest emigration. Increased expression of *Cv-2* caused premature onset of trunk neural crest cell migration in the chick embryo indicative of Cv-2 acting to promote BMP activity (Coles et al. 2004).

How are Wnts involved in the process of delamination? Recent work from the Kalcheim group showed that the premature emigration of neural crest precursors from the caudal neural tube after down-regulation of *Noggin* is accompanied by an up-regulation of *Wnt1* in the dorsal neural tube. They showed also that while *Noggin*-expressing cells are able to inhibit *Wnt1* expression in the dorsal neural tube, grafts of BMP-4-coated beads on the dorsal neural tube at the level of segmental plate resulted in premature up-regulation of *Wnt1* mRNA. Furthermore, inhibition of the canonical pathway prevented neural crest delamination while the over-expression of β-catenin rescued neural crest delamination in Noggin-inhibited neural primordia. Thus, in the delamination process it could be concluded that BMP-dependent Wnt signalling is necessary for the EMT of neural crest cells (Burstyn-Cohen et al. 2004).

As mentioned previously, *Slug*, one of the earliest known neural crest markers, has a functional Lef/β-catenin binding site on its promoter, implying that its expression is under the control of Wnt signalling. When *Slug* expression was inhibited with antisense oligonucleotides in chicken embryos after neural crest induction, neural crest cells positive for HNK-1 were present but failed to migrate out of the neural tube (Nieto et al. 1994). This suggested that *Slug* is required for neural crest delamination. Furthermore, in *Xenopus*, transplantation of neural folds in which *Slug* expression was inhibited with a Slug antisense oligonucleotide into normal embryos showed no neural crest migration. This suggests also that cell autonomous *Slug* expression is important for EMT (Carl et al. 1999). In the chick neural tube the over-expression of *Slug* resulted in an increase in neural crest production. Slug itself was able to induce the expression of *RhoB*, which has been shown to play a role in the delamination process of neural crest cells (Liu and Jessell 1998), leading to an increase in the number of HNK-1-positive migratory cells. It has to be mentioned that whereas in the cranial region *Slug* increased the number of premigratory and migratory neural crest, in the spinal cord *Slug* was only able to increase the number of premigratory crest precursors (Del Barrio and Nieto 2002). Consistent with the results from Nieto et al. (1994) in chicks, it was shown in *Xenopus* embryos when a dominant-negative form of *Slug* was expressed under hormone-inducible control, an early inhibition of Slug activity blocked the formation of neural crest, while a late inhibition of Slug, after neural crest cell induction, blocked the delamination of the neural crest cells from the dorsal neural tube (La Bonne and Bronner-Fraser 2000). Thus, the induction of *Slug* expression is an important intermediatory step in the ability of Wnts to induce the delamination and migration of the neural crest cells.

1.4
Outline of Research Plan

1.4.1
Wnt6 and Somitogenesis

Somites are epithelial structures that develop in the paraxial mesoderm on both sides of the neural tube/notochord in an anterior-to-posterior direction. The epithelial organisation of somites is crucial for the later morphogenesis of trunk muscles. Paraxis mutants are unable to form epithelial somites, and although the specification of skeletal muscles takes place, their organisation is altered very strongly (Burgess et al. 1996; Wilson-Rawls et al. 1999). These data indicate that muscle progenitors use the epithelial organisation displayed in somites and in dermomyotome as a scaffold on which to develop. Work of our own group has shown that the expression of *Paraxis* is dependent on factors produced in the ectoderm (Sosic et al. 1997). After ablating the overlying ectoderm at the level of the caudal segmental plate no somites form at the operation level. The expression of *Paraxis* in these embryos is diminished, showing the dependence of *Paraxis* expression on the maintenance of an intact surface ectoderm. We therefore want to identify the factor produced by the ectoderm which is able to induce *Paraxis* expression in the paraxial mesoderm thereby promoting the epithelialisation leading to somite formation. *Wnt6* is the only Wnt known to be expressed in the ectoderm overlying the segmental plate and epithelial somites in the chick embryo (Cauthen et al. 2001; Schubert et al. 2002; Rodriguez-Niedenführ et al. 2003), which makes it a good candidate to play a role in the epithelialisation process. We therefore propose that Wnt6 is the epithelialisation factor necessary for somite formation in the chick.

To test this hypothesis, we performed different surgical manipulations including separation experiments, the implantation of *Wnt6*-expressing cells and the usage of the Wnt antagonist Sfrp2.

1.4.2
The Regulation of *cNkd1* Expression

Nkd is one of the target genes of the Wnt signalling cascade, whose activity is required to restrict Wnt signalling during segmentation in *Drosophila* development, thereby generating a negative feed-back loop. Nkd is not only able to inhibit the canonical Wnt pathway but also to activate the planar cell polarity (PCP) pathway, suggesting that Nkd may act as a switch to direct Dsh activity towards the PCP pathway (Yan et al. 2001). We wanted to examine the expression profile of *Nkd1* in the chick to determine whether there are similarities to the expression described in the mouse (Ishikawa et al. 2004). Furthermore, we were interested in the regulative tissues and factors involved in *cNkd1* expression. After cloning *cNkd1*, we therefore examined its normal expression and performed a set of different surgical manipulations including separation experiments and the implantation of different Wnt expressing cells to enlighten the regulation of *cNkd1*.

1.4.3 Wnt6 and Neural Crest

The formation and distribution of neural crest cells includes two important steps: (1) induction and (2) delamination, which is followed by migration of these cells to their final destination. Neural crest cells have to undergo an epithelio-mesenchymal transition after their induction to move away from the neural tissue. Interactions between non-neural ectoderm and naïve neural plate tissue are required to induce the formation of neural crest cells (Selleck and Bronner-Fraser 1995; Dickinson et al. 1995). We wanted to determine if Wnt6, as the only known Wnt in the chick surface ectoderm, could be the ectodermal Wnt involved in the formation of neural crest cells. To test this hypothesis we performed a complex set of surgical manipulations and cell culture experiments including neural crest in vitro cultures, implantation of *Wnt* expressing cells, electroporation of constructs of the different Wnt signalling pathways and the employment of short interfering RNAs (siRNA) against Wnt6.

2 Materials and Methods

2.1 Embryos

Fertilised eggs were obtained from Lohmann Tierzucht (Germany) and Henry Stewart and Co. (UK) and incubated at 38°C up to HH10–12 (Hamburger and Hamilton 1951). In HH12 embryos the unsegmented paraxial mesoderm reaches its maximum length; it is therefore the best time for manipulations influencing somite formation.

2.2 Culture of Wnt-Expressing Cells

Wnt-expressing murine NIH 3T3 cells and control cells (NIH 3T3 cells transfected with empty LacZ vector) were cultured in DMEM GlutaMax I (Invitrogen), 10% fetal calf serum (FCS), 1% penicillin/streptomycin, 1% sodium pyruvate, 0.025% G418. Transfected cells were grown at 37°C, 5% CO_2 until confluent.

2.3 Operations

All operations were performed at the level of the unsegmented paraxial mesoderm (segmental plate) on 2-day-old embryos (HH10–12).

1. Ectoderm removal and replacement with gold foil. Ectoderm destined for ablation was first marked using Nile blue sulphate and was non-enzymatically removed using a tungsten needle.
2. Implantation of Wnt-producing cells beneath the ectoderm, medial or lateral to the paraxial mesoderm. Care was taken to apply cells directly under the ectoderm using a micropipette without disturbing contact between unoperated tissues. Importantly, the ectoderm at the level of cell implantation was never damaged.
3. Ectoderm removal, replacement with gold foil and implantation of Wnt-producing cells beneath the gold foil. This was a combination of procedures 1 and 2.
4. Implantation of Wnt-producing cells beneath the ectoderm and of an impermeable polycarbonate membrane between the neural tube/notochord and the segmental plate.
5. Implantation of control beads (PBS beads) or Sfrp2 beads beneath the ectoderm at the level of the segmental plate. Affigel blue beads were soaked in PBS or Sfrp2 (R and D Systems, USA) (2 μg/ul, made in PBS) for a minimum of 12 h.
6. Implantation of an impermeable polycarbonate membrane between the neural tube and the segmental plate.
7. Extirpation of the neural tube or notochord at the level of the cranial segmental plate.
8. Implantation of cells into the neural tube at the level of the caudal segmental plate at HH10–12 using fine pulled glass micropipettes. When the injections were performed in HH10 embryos, cells were implanted before the neural tube had closed. In some cases, this affected the closure of the neural tube; however, there were no differences in the results obtained from embryos that had closed neural tubes compared to those that had not.
9. Electroporation of the neural tube was performed as described by Nakamura and Funahashi (2001). The DNA solution was injected in the neural tube at the appropriate level and electrodes were placed next to the embryo at the level of DNA injection. By using an Intracell (Royston, UK) TTS pulser, five square pulses of 10 mV, 20 ms apart were applied. All embryos were electroporated into one side of the neural tube at the level of the caudal segmental plate at HH12. Constructs containing Δ-DIX-Dvl, an inhibitor of canonical Wnt signalling and Δ-DEP-Dvl, an inhibitor of non-canonical signalling are as described in Rosso et al. (2005). The dominant-negative TCF-4 construct in pCDNA3 vector was as described in Tetsu and McCormick (1999). The constitutively active β-catenin construct in CS105 vector was as described in Baker et al. (1999). A full-length chick Naked cuticle 1 (cNkd1) gene was cloned into RCAS (Schmidt et al. 2006).

2.4
BrdU Experiment

Embryos were incubated in ovo for 30 min with 300 μl of a 16 mM solution of BrdU (Sigma). Cells incorporating BrdU were detected using mouse anti-BrdU antibody (Roche) 1:100 and a secondary goat anti-mouse alkaline phosphatase (Dako), 1:200, developed with NBT/BCIP (Roche) (see also Sect. 2.8). Sections were photographed on the Leica DMRA2 microscope (Leica Microsystems, Wetzlar, Germany).

2.5
Neural Crest Cell Culture

Neural tubes from the level of the segmental plate of HH10 embryos were dissected in ice-cold Hanks buffer, allowed to recover in L15 medium/10% FCS and placed onto confluent layers of Wnt-expressing cells or control cells. The neural crest cells were allowed to migrate out from the explanted neural tubes for 20–24 h. They were subsequently labelled with mouse HNK-1 antibody (Serotec, Oxford, UK). The numbers of cells in eight separate cultures were counted double-blind and analysed statistically and plotted graphically in GraphPad Prism software.

2.6
HNK Staining Neural Crest Culture

Cultures were fixed in paraformaldehyde (PFA) 4% in phosphate-buffered solution (PBS) overnight, washed in PBT (PBS with 0.05% Triton) 0.05% for 2×5 min, incubated in 0.1% H_2O_2 in PBT 0.05% for 1 h, washed in PBT 0.05% for 5 min and blocked in 10% goat serum in PBT 0.05% for 30 min. After this the cultures were incubated with mouse HNK-1 antibody (Serotec) 1:100 for 2 h. Cultures were washed 3×10 min in 10% goat serum in PBS and incubated with the secondary anti-mouse-AP antibody (Dako) 1:100 in 10% goat serum in PBS for 45 min. Then cultures were washed 1×5 min in PBS, 1×5 min in AP buffer and developed using NBT/BCIP (Roche).

2.7
Wnt6 siRNA RCAS Construct

Wnt6 target sequences were selected using Sfold (https://sfold.wadsworth.org/sirna.pl) and Genscript (https://www.genscript.com/ssl-bn/app/rnai), based upon their stability (Δ E). Two target sequences, T1 (Δ E 12.3) and T2 (Δ E 11.94), were selected and forward and reverse primers designed for each [T1=acgacgtgcagtttgg ctatga, T2 = cactcattgatctgcacaacaa]. These primers were used along with flanking primers to generate the siRNA construct, which was cloned into the pRFPNAiC vector containing the chick U6 promoter. This was directionally subcloned into a modified RCAS vector (see Das et al. 2006 for more details). DF-1 cells were infected with the construct and supernatant harvested after 7 days. The supernatant was concentrated by centrifugation to give a minimum titre of 10^8 plaque-forming units per millilitre. This was injected onto the surface of the right side of the developing embryo at HH7–11, from the level of the last 2–3 somites to the caudal tail bud. Embryos were allowed to develop for either 2 or 3 days. Control was empty vector.

2.8 Immunohistochemistry on Sections

Embryos were fixed in 4% PFA and processed for either wax or frozen sectioning. Wax sections were blocked for 30 min with 20% goat serum in PBS before incubation with mouse HNK-1 (Serotec) 1:100 or mouse anti-BrdU 1:200 (Roche) dilution overnight at 4°C, washed 3×10 min in PBS, followed by goat anti-mouse-AP secondary (Dako) 1:200 dilution for 1 h at 20°C, washed 2×10 min in AP buffer, and developed using the NBT/BCIP mix (Roche). Other sections were incubated with mouse phosphorylated Jun kinase (pJNK) (Santa Cruz) 1:200 or mouse Wnt6 (Zymed, UK) 1:50 dilution followed by standard ABC secondary-horse-radish peroxidase (HRP) antibody labelling (Vectastain, Vector Laboratories, Peterborough, UK) and development with diaminobenzidine (DAB). Sections were photographed on a Leica DMRA2 microscope and captured using IM500 image manager software. Frozen sections were blocked for 30 min with 20% goat serum in PBS before incubation with mouse de-phosphorylated (active) β-catenin (Upstate) 1:50, mouse N-cadherin (Zymed) 1:100, or mouse Laminin (Sigma) 1:100 dilution overnight at 4°C, washed 3×10 min in PBS, followed by goat anti-mouse Alexa-488 (Molecular Probes), 1:200 dilution. The sections incubated with antibody against active β-catenin were mounted using DakoCytomation fluorescent mounting media (S3023) containing 7.5 μl DAPI (Molecular Probes) solution (5 mg/ml). Sections were imaged on a Zeiss LSM510 confocal microscope.

2.9 Immunohistochemistry on Whole Embryos

Whole embryos were fixed in 4% PFA overnight, washed in PBS 3×1 h, and incubated with 0.05% hydrogen peroxide in PBT overnight at 4°C. After washing in PBT for 2×1 h they were incubated in mouse neurofilament antibody RMO-270 (Cambridge Biosciences) 1:10,000 in PBT overnight at 4°C. The antibody solution was removed and the embryos were washed 2×1 h in PBT containing 1% goat serum. The embryos were incubated with the secondary anti-mouse-AP antibody (Dako) 1:200 in PBT/1% goat serum overnight at 4°C. After washing in PBT for 1×30 min, and washing in AP buffer for 2×30 min, the embryos were developed using NBT/ BCIP (Roche).

2.10 Whole-Mount In Situ Hybridisation

Embryos were washed in PBS and then fixed overnight in 4% PFA at 4°C. The antisense RNA probe was labelled with digoxigenin and whole mount in situ hybridisation was performed as described by Nieto et al. (1996).

2.11 Chemicals

- PBS (phosphate buffered saline): prepare using Dubecco PBS tablets, Oxoid
- PFA (paraformaldehyde), Sigma
- Glutaraldehyde, Sigma
- Diethyl-pyrocarbonate (DEPC), Sigma
- NaCl, Sigma
- Sodium citrate, Sigma
- $MgCl_2$, Sigma
- Tris base, Sigma
- HCl, Merck
- NaOH, Sigma
- Triton X-100, Sigma
- Tween, Sigma
- CHAPS, Sigma
- NBT, Roche
- BCIP, Roche
- Proteinase K, Roche
- Anti-DIG Antibody, Roche
- Goat serum, Invitrogen
- Methanol, Merck
- Formamide, 250 ml, Fluka
- Blocking powder, 50 g, Roche
- tRNA from yeast, Roche (cat number 109 223)
- EDTA 0.5 M pH 8.0, Roche
- Heparin lithium salt, 100,000 Units, Sigma

2.12 Anti-sense RNA Probes

The following probes were used in this study: *Pax3*, 645-bp fragment corresponding to nucleotides 468–1,113 and *Pax1* (gift from Dr. Martin Goulding), *Paraxis*, full-length clone (see Sosic et al. 1997) was a gift from Professor Eric Olson. *MyoD*, clone CMD9 full-length, 1.5-kb, fragment (gift from Professor Bruce Patterson), *Sox10, FoxD3, RhoA, RhoB, MafB, Cad 6B* (gift from A. Graham), *Slug* (gift from A. Nieto), cNkd1 (cloned in our lab). The embryos were cryosectioned for further examination.

Gene	Size (kb)	Linearisation enzyme	Transcription enzyme
Pax3	0.645	Not I	T3
Paraxis	1.38	BamH I	T7
Pax1	0.4	EcoRV	T7
MyoD	1.5	Sac I	T3
cNkd1	0.542	EcoR I	T7
MafB	1.0	Hind III	T3
RhoA	1.2	Not I	T3
RhoB	2.4	Not I	T7
Sox10	1.0	EcoRI	T3
FoxD3	1.4	Xho I	T3
Cad6B	1.0	Hinc II	T3
Slug	1.8	Xba I	T7

2.12.1 DNA Linearisation

- 20 μg DNA; X μl
- 10×Restriction buffer; 10 μl
- DEPC water; Add to 100 μl
- Enzyme (10–50 U/μl); 5 μl
  - Incubate at appropriate temperature (according the used enzyme) for 3 h
  - Check linearisation by running 1 μl on 1% gel

2.12.2 DNA Purification

- Take the linearised DNA and add equal volume of phenol/chloroform
- Vortex properly 2 min
- Centrifuge (13000 rpm) for 10 min
- Save the top layer, discard the bottom layer
- Add to the top layer equal volume of chloroform
- Vortex 2 min
- Centrifuge (13000 rpm) for 10 min
- Save the top layer, discard the bottom layer
- Add 1 μl glycogen solution (10 mg/ml) and 1/10 volume of 3 M sodium acetate pH 5.2
- Vortex by hand 10 s
- Add 2.5 volumes of ethanol
- Vortex by hand 10 s
- Place on dry ice for minimum of 30 min
- Centrifuge (13000 rpm) for 10 min at 4°C
- Wash precipitate in 200 μl ethanol 70%

- Vortex properly 60 s
- Centrifuge(13000 rpm) for 10 min at room temperature
- Re-suspend pellet in 20 μl DEPC water
- Run 1 μl on 1% gel to estimate concentration

2.12.3
Probe Synthesis

- DEPC water; add to 20 μl
- 5×Transcription buffer; 4 μl
- 100 mM DTT; 2 μl
- DNA; volume that contains 1 μg
- RNase inhibitor; 0.5 μl
- DIG RNA labelling mix; 2 μl
- Transcriptase (T3, T7 or Sp6); 2 μl
  - Incubate at 37°C overnight
  - Add 80 μl DEPC water and pull through spin columns to remove un-incorporated nucleotides
  - Run 5 μl on 1% gel to estimate concentration
  - Store at −80°C
  - Use 50–100 ng of probe for every millilitre of hybridisation solution

3
Results

3.1
Wnt6 and Somitogenesis

3.1.1
Consequences of Ectoderm Removal on the Epithelialisation of Segmental Plate

In all cases operations were performed at the level of the caudal segmental plate. Ectoderm removal resulted in abolition of somite formation on the operated side. *Pax3* is a useful marker since it is expressed at high levels in the epithelialised dorsal compartment of the somite.

Ectoderm removal followed by re-incubation resulted in a dramatic down-regulation in the high levels of *Pax3* expression associated with epithelialised dermomyotome at the site of operation (Fig. 3A). Some *Pax3* expression was detected directly adjacent to the dorsal neural tube coinciding with a small region that remained in an epithelial state. Examination of a marker of the ventral compartment, *Pax1*, following ectoderm removal showed robust expression of the gene at the site of operation. Expression levels of *Pax1* on the operated side were similar to those on the control side (Fig. 3B). *Paraxis*, a marker for the epithelialisation of the paraxial mesoderm (Burgess et al. 1995; Barnes et al. 1997; Sosic et al. 1997), was down-regulated like *Pax3* at the site of ectoderm removal (Fig. 3C). No evidence of cell death following ectoderm removal could be detected (Schmidt et al. 1998).

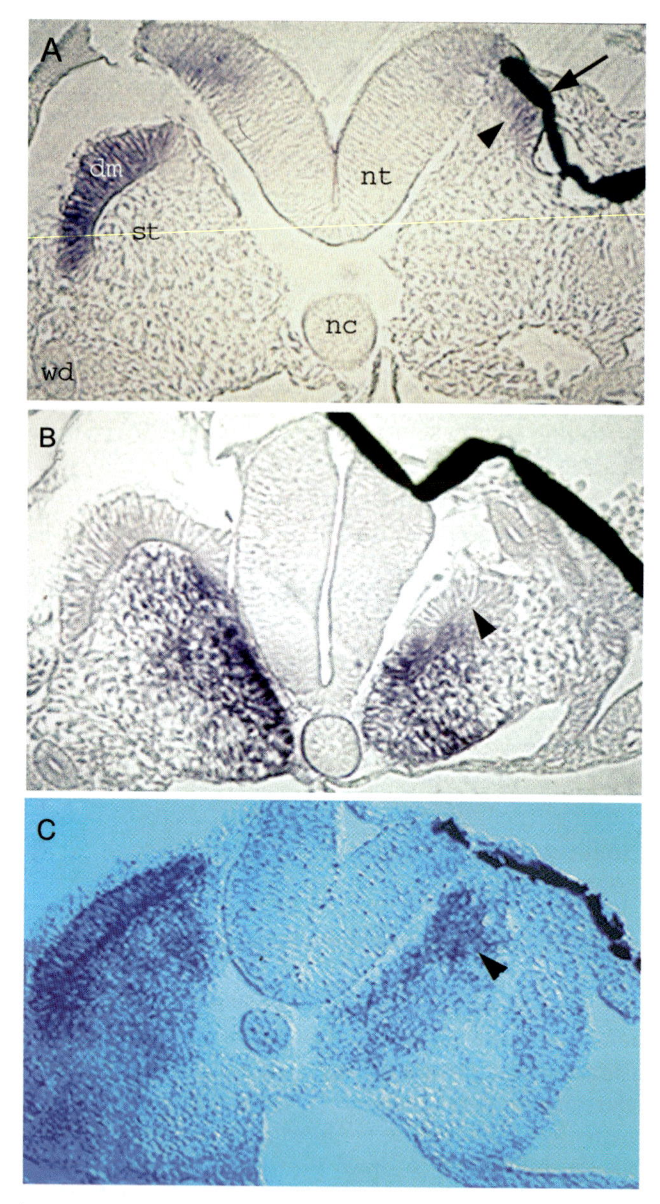

Fig. 3 A–C Ectoderm removal prevents somite formation. Transverse section of a 3-day-old chick embryo after ectoderm removal and replacement by gold foil showing expression of **A** *Pax3*, **B** *Pax1 and* **C** *Paraxis*. **A** On the operation side, only the medial lip of the dermomyotome (*arrowhead*) near to the dorsal neural tube shows strong *Pax3* expression. *dm*, dermomyotome; *st*, sclerotome; *wd*, wolffian duct; *nc*, notochord; *nt*, neural tube; *arrow* marks the implanted gold foil. **B** The *Pax1* expression on the operation side is not changed in comparison to the control side. The *arrowhead* marks the medial lip of the dermomyotome. **C** *Arrowhead* marks the weak *Paraxis* expression on the operation side; there is no dermomyotome

Ectoderm removal therefore results in the loss of the epithelial organisation of the somites except in a small region directly adjacent to the dorsal neural tube. At the molecular level, removal of the ectoderm leads to the loss of epithelial somite markers, except in the region adjacent to the neural tube. This, however, is not accompanied by an up-regulation of sclerotome markers.

3.1.2
Effect of Wnt6 on the Development of Somites

Wnt6 is a good candidate as an epithelialising factor originating from the ectoderm as it is not only expressed in this tissue but in other systems too, where it has been shown to confer epithelialising activity (Kispert et al. 1998; Itaeranta et al. 2002). We investigated the effect of Wnt6 on the development of somites by implanting Wnt6 protein-expressing cells at the caudal segmental plate level but taking care to leave the ectoderm in place. Implantation of Wnt6-producing cells led to the strong expression of *Pax3* at the site of operation (Fig. 4A). The site of operation was noted at the end of each procedure. Wnt6-expressing and control cells could be easily identified after fixation since they appear as a white stripe. Transverse sections showed on the un-operated side a differentiated somite with a *Pax3*-positive epithelial dermomyotome and a *Pax3*-negative mesenchymally organised sclerotome. On the operated side. however, we detected the presence of an epithelial somite (Fig. 4B). Furthermore, somites on the operated side were smaller compared to those on the un-operated side (Fig. 4A and B). Expression of *Pax3* at the site of operation resembled that found in normally occurring epithelial somites. *Wnt6* expression also caused a decrease in *Pax1* expression (Fig. 4D and E).

Expression of *Paraxis* on the operated side was similar to the expression of *Pax3*, reminiscent of the predicted expression pattern of this gene in epithelial somites (Fig. 4F and G). The application of *Wnt6*-expressing cells beneath the ectoderm dorsal to the segmental plate led to the development of two rows of epithelial somites (Fig. 4H and I). We therefore subsequently examined the effect of Wnt6 on the myogenic programme. Expression of *MyoD* was transiently delayed on the operated side (Fig. 4J). The expression domain was smaller than the un-operated side (Fig. 4L) and this leads to a compact domain of expression (Fig. 4K). We can therefore summarise that the over-expression of *Wnt6* prolongs the epithelial state of the somites. This results in a delay in somite differentiation and a decrease of muscle precursor migration signified by a reduction in *Pax3* expression in the limb (Fig. 4C).

3.1.3
Wnt6 Rescues the Effect of Ectoderm Removal

Since Wnt6 maintains epithelialisation of somites we investigated whether this protein could mimic the effect of the dorsal ectoderm. To this end, we removed the ectoderm and implanted Wnt6-producing cells to the caudal segmental plate. Whereas ectoderm removal alone leads to the loss of epithelial somite formation and high

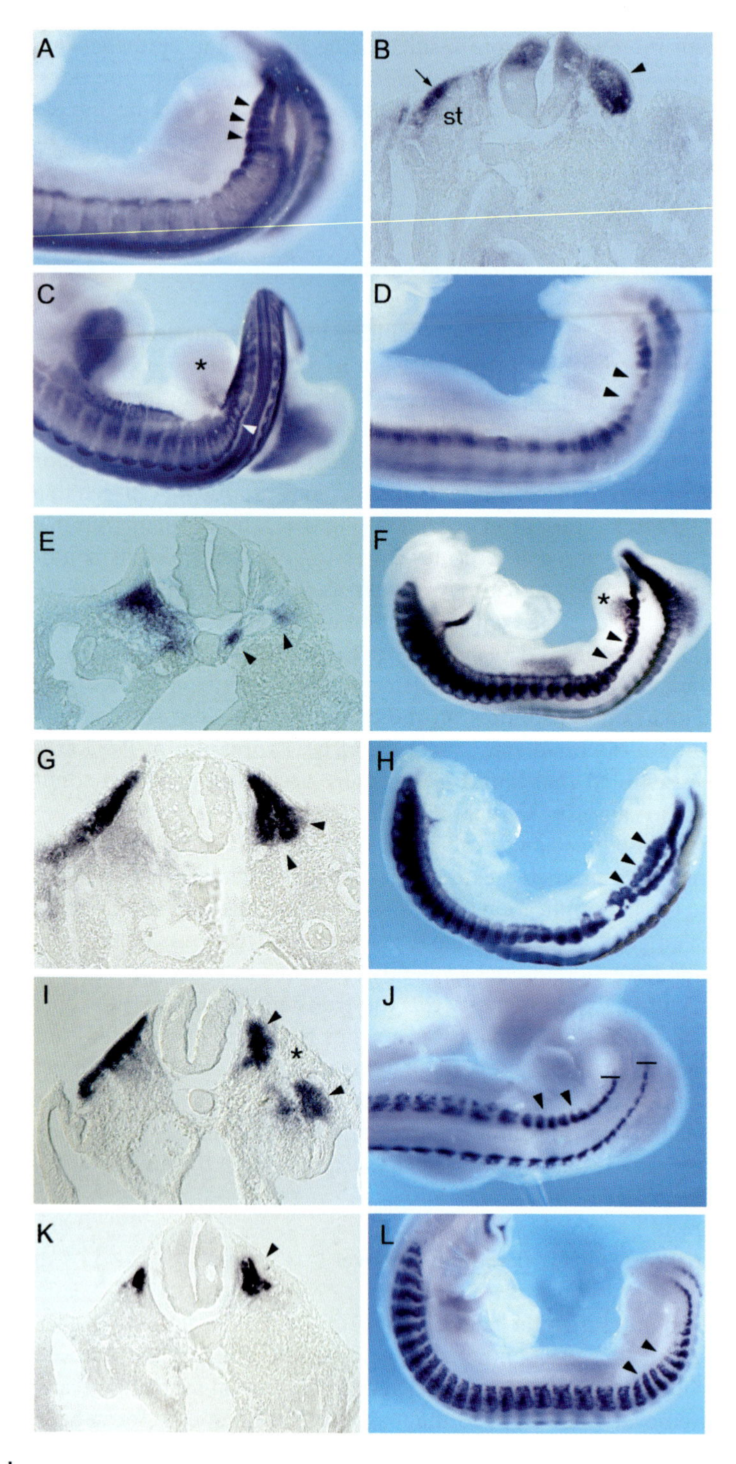

Fig. 4 A–L

levels of *Pax3* expression, additional implantation of *Wnt6*-expressing cells rescued not only the expression of *Pax3* but also the epithelialisation of the segmental plate (Fig. 5A). In a similar manner, Wnt6-producing cells rescued expression of *Paraxis* (Fig. 5B). *MyoD* expression was delayed following the dual procedure (Fig. 5C).

3.1.4 Wnt6 Rescues the Effect of Medial Barrier Insertion

Insertion of an impermeable barrier between the neural tube/notochord and the caudal segmental plate leads to the loss of epithelial somites and the expression of genes including *Pax3* and *Paraxis*. We investigated whether Wnt6 could rescue the formation of somites and expression of these genes following medial tissue isolation. Implantation of a medial barrier and Wnt6-producing cells at the level of

Fig. 4 A–L Wnt6 induces epithelialisation of paraxial mesoderm. **A** Whole mount in situ hybridisation for *Pax3* of a 3-day-old chicken embryo after implantation of Wnt6-expressing cells lateral to the segmental plate. Stronger *Pax3* expression is found on the operation side (marked by *arrowheads*). **B** Transverse section of the embryo shown in **A**. On the operation side an epithelial somite is found (marked by an *arrowhead*) which is *Pax3* positive; on the control side the somite is differentiated into the dermomyotome (marked by an *arrow*) and the sclerotome (*st*). **C** Whole mount in situ hybridisation for *Pax3* of a 3-day-old chicken embryo after implantation of Wnt6-expressing cells beneath the ectoderm at the level of segmental plate. On the operation side the somites are smaller (marked by an *arrowhead*) and *Pax3* expression in the limb bud is reduced (*asterisk*). **D** Whole mount in situ hybridisation for *Pax1* of a 3-day-old chicken embryo after implantation of Wnt6-expressing cells lateral to the segmental plate. On the operation side (*arrowheads*) the *Pax1* expression is reduced. **E** Transverse section of the embryo shown in **D**. *Pax1* expression (*arrowheads*) is markedly reduced on the operation side compared to the control side. **F** Whole mount in situ hybridisation for *Paraxis* of a 3-day-old embryo after implantation of Wnt6-expressing cells lateral to the segmental plate. On the operation side there are smaller but increased number of somites (*arrowheads*). *Paraxis* expression is reduced in the limb bud (*asterisk*). **G** Transverse section of the embryo shown in **F**. On the operation side a somite-like *Paraxis*-positive structure is found (*arrowheads*) whereas on the control side the somite is differentiated into the dermomyotome and sclerotome. **H** Whole mount in situ hybridisation for *Paraxis* of a 3-day-old chicken embryo after implantation of Wnt6-expressing cells beneath the ectoderm at the level of segmental plate. The operation side shows two rows of *Paraxis*-positive somites. **I** Transverse section of the embryo shown in **H**. On the operation side two *Paraxis*-positive epithelial somites are found (*asterisk* marks implanted Wnt6-expressing cells). On the control side the somite is differentiated in dermomyotome and sclerotome. **J** Whole mount in situ hybridisation for *MyoD* of a 3-day-old chicken embryo. On the operation side the somites are smaller; *MyoD* expression seems more condensed. The *MyoD* expression itself is retarded on the operation side in comparison to the control side (levels of *MyoD* expression on operation and control side are marked by *black lines*). **K** Transverse section of the embryo shown in **J**. On the operation side the *MyoD* expression is more condensed (*arrowhead*). **L** Whole mount in situ hybridisation for *MyoD* of a 3-day-old chicken embryo after implantation of Wnt6-expressing cells lateral to the segmental plate. The somites on the operation side are smaller (marked by *arrowheads*)

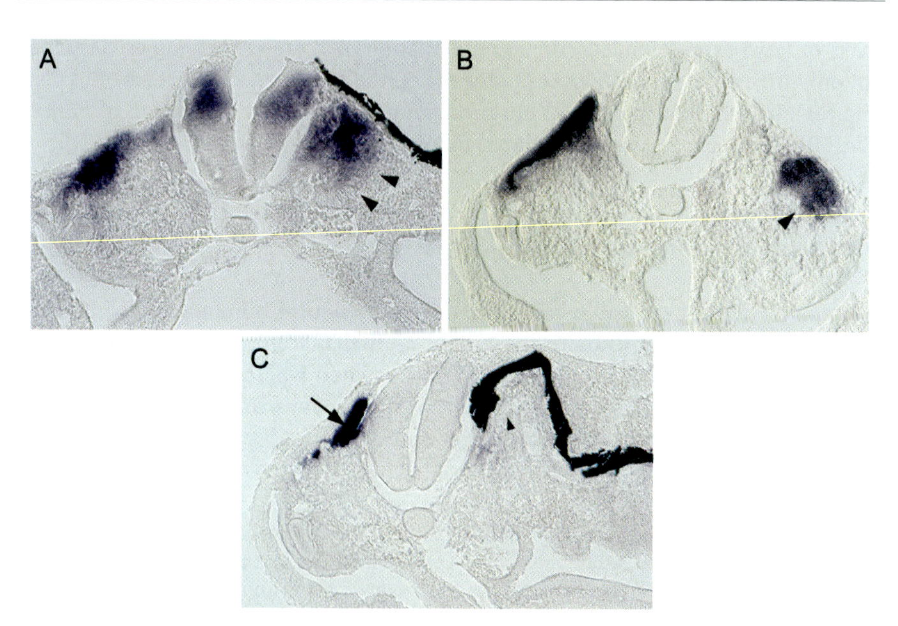

Fig. 5 A–C Wnt6 rescues somite formation after ectoderm removal. Transverse section of a 3-day-old chick embryo after ectoderm removal and implantation of Wnt6-expressing cells beneath the gold foil and following re-incubation for 18 h showing expression of **A** *Pax3*, **B** *Paraxis* and **C** *MyoD*. **A** On the operation side the Wnt6-expressing cells enable the epithelialisation of the paraxial mesoderm and initiate strong *Pax3* expression. *Arrowheads* mark the *Pax3*-positive somite on the operation side. **B** On the operation side a somite expressing *Paraxis* is found (*arrowhead* marks the somite). Note that the gold foil has been removed during processing. **C** On the operation side no *MyoD* expression is found (*arrowhead* marks the epithelial structure in the paraxial mesoderm induced by the Wnt6-expressing cells). The control side shows normal *MyoD* expression in the myotome (*arrow*)

Fig. 6 A–H (continued) **C** On the operation side *Paraxis* expression is found lateral to the polycarbonate membrane (marked by *arrowheads*). **D** Transverse section of the embryo shown in **C**. On the operation side a somite-like structure expressing *Paraxis* is found lateral to the polycarbonate membrane (*arrowheads*). **E–H** Whole mount in situ hybridisation of a 3-day-old chicken embryo after implantation of Lac-Z (control) cells beneath the ectoderm of the segmental plate and additional implantation of a polycarbonate membrane between neural tube/notochord and segmental plate showing expression of *Pax3* (**E**, **F**) and *Paraxis* (**G**, **H**). **E** On the operation side no epithelial structures and no *Pax3* expression are found (marked by *arrowheads*). **F** Transverse section of the embryo shown in **E**. On the operation side no *Pax3* expression is found lateral to the polycarbonate membrane (*arrowheads*). **G** On the operation side no *Paraxis* expression is found lateral to the membrane. **H** Transverse section of the embryo shown in **G**. On the operation side no *Paraxis* expression is found lateral to the polycarbonate membrane

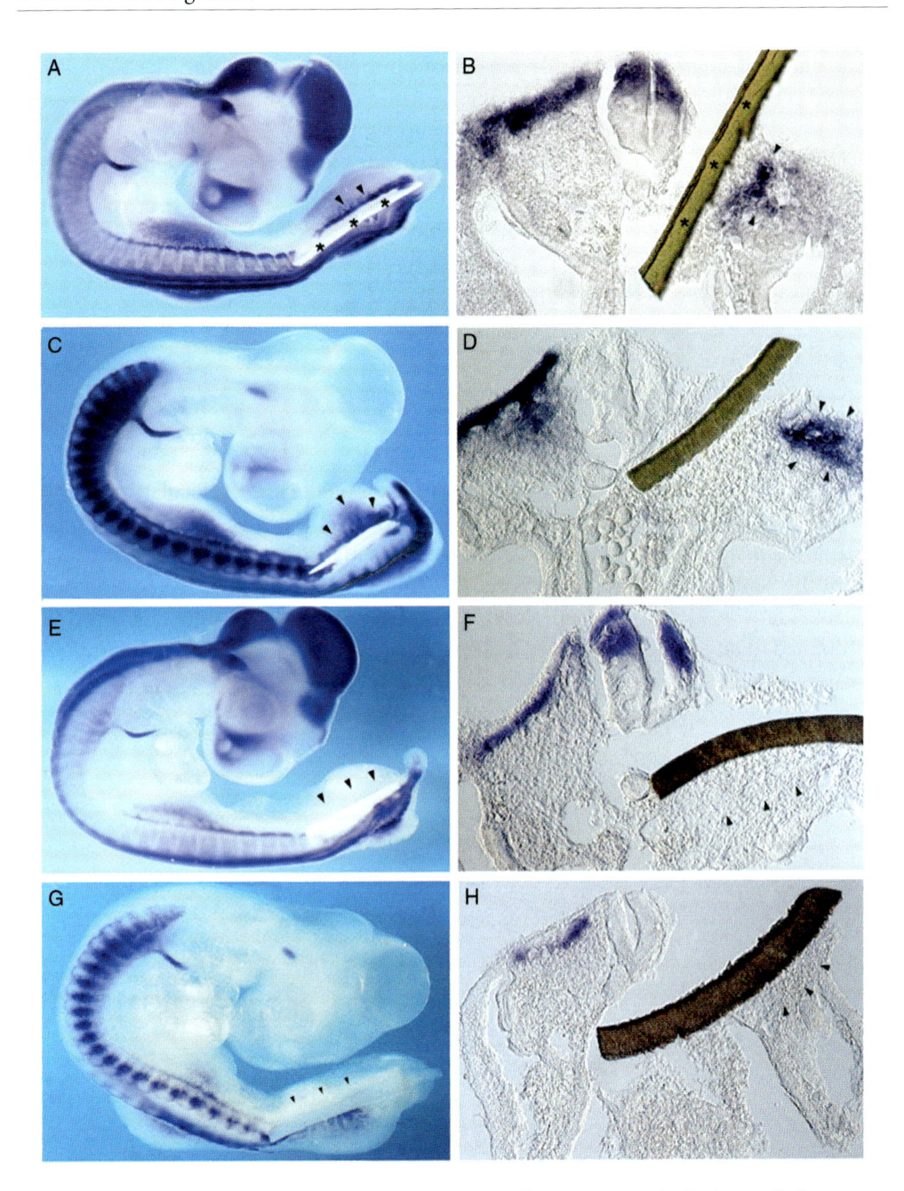

Fig. 6 A–H Wnt6 rescues somite formation after medial structure isolation. Whole mount in situ hybridisation of a 3-day-old chicken embryo after implantation of *Wnt6*-expressing cells beneath the ectoderm of the segmental plate and additional implantation of a polycarbonate membrane (marked by *asterisks*) between neural tube/notochord and segmental plate showing expression of *Pax3* (**A**, **B**) and *Paraxis* (**C**, **D**). **A** On the operation side *Pax3* expression is found near to the implanted cells (*arrowheads*) and lateral to the implanted polycarbonate membrane (marked by *asterisks*). **B** Transverse section of the embryo shown in **A**. On the operation side *Pax3* expression is found in the paraxial mesoderm (marked by *arrowheads*) lateral to the polycarbonate membrane (marked by *asterisks*)

the segmental plate led to the formation of epithelial somites and the expression of *Pax3* (Fig. 6A and B) and *Paraxis* (Fig. 6C and D). After implantation of a medial barrier and LacZ-expressing cells, there was no sign of rescue of either epithelialisation or the expression of *Pax3* (Fig. 6E and F) or *Paraxis* (Fig. 6G and H). There was no evidence of cell death following medial barrier insertion (Schmidt et al. 1998). These experiments show a signal responsible for the epithelialisation of the somites originates from medial structures. *Wnt6* over-expression is able to rescue the effect of the loss of medial signalling.

3.1.5 Effect of Wnt6 Expression on Muscle Development

Examination of embryos treated with Wnt6-producing cells with an intact ectoderm led to a significant decrease in the expression of *MyoD* in the limb musculature (Fig. 7A) compared to the control site at same level (Fig. 7B). Transverse section of the limb on the operation side showed a significant decrease of *MyoD* expression and the massive loss of the ventral and dorsal muscle masses (Fig. 7C) compared to the section of the control side (Fig. 7D) at the same level. These results show that cells that were once maintained in an epithelial state are eventually capable of muscle differentiation.

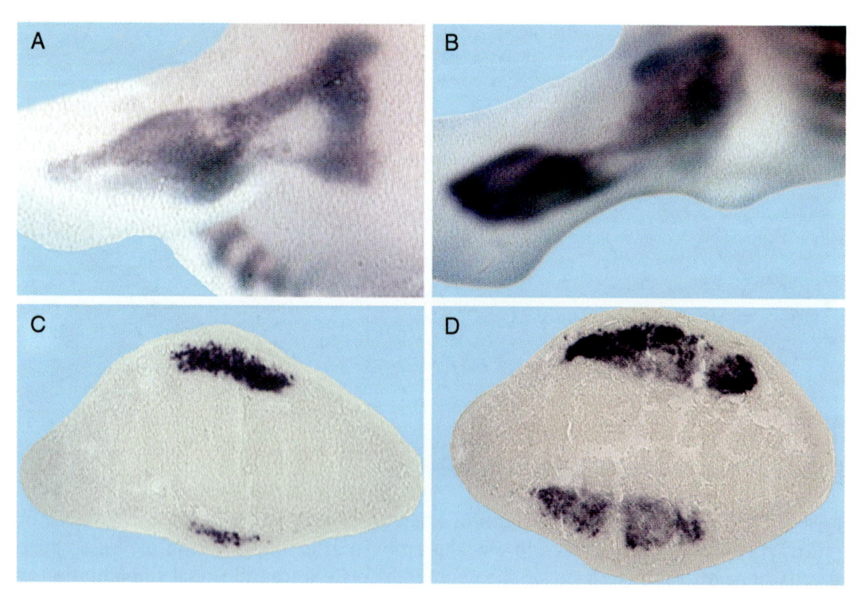

Fig. 7 A–D Wnt6 leads to decreased limb muscle development. **A** Whole mount in situ hybridisation for *MyoD* of a stage-26 embryo after implantation of *Wnt6*-expressing cells beneath the ectoderm at the level of segmental plate. The lower limb of the operation side is shown. The *MyoD* expression compared to the control side shown in **B** is markedly reduced. **B** The lower limb of the control side with normal *MyoD* expression. **C** Transverse section of the limb shown in **A**. The loss of the myogenic content in both dorsal and ventral muscle mass is evident. **D** Transverse section of the control limb shown in **B**

3.1.6
Effect of Sfrp2 on Somite Development

Experiments outlined thus far suggest that Wnts play a role in the epithelialisation process. In order to investigate this possibility we tested the hypothesis that application of Wnt antagonists to the segmental plate should prevent the epithelialisation of somites. Secreted Fz-related proteins (Sfrps) are potent antagonists of Wnt signalling and are expressed in a dynamic manner in the segmental plate and somites (Hoang et al. 1998; Lee et al. 2000). Implantation of Affigel beads soaked with Sfrp2 protein at a concentration of 2 μg/μl at the level of the caudal segmental plate led to the inhibition of epithelialisation of the segmental plate. Either no somites were formed (Fig. 8A and B) or very small somites developed directly in contact with the neural tube (Fig. 8C and D). Expression of *Paraxis* was either completely abolished at the site of Sfrp2 bead implantation (Fig. 8A and B) or was found in the small somites adjacent to the neural tube (Fig. 8C and D). In both cases, however, expression of *Paraxis* as an epithelial marker was reduced at the operation site. Implantation of Sfrp2 beads at the level of cranial segmental plate led to formation of smaller somites that expressed *Pax3* (Fig. 8E, arrowheads). Implantation of Sfrp2 beads at the caudal segmental plate resulted in the loss of the epithelialisation of segmental plate, which displayed a *Pax3* expression pattern similar to its profile in unsegmented paraxial mesoderm (Fig. 8E, arrows). Robust expression of *Pax3* following Sfrp2 bead application also indicated that apoptosis was not induced. Implantation of control beads at the level of segmental plate had no effect on somite formation (Fig. 8F).

3.2
The Regulation of *cNkd1* Expression

3.2.1
Normal Expression of *cNkd1* in Chicken Embryos from HH8–HH20

In situ hybridisation was used to determine the expression profile of the cloned *cNkd1*. Strong expression was not detected in embryos younger than HH8. At HH8, expression was detected at three major sites: (1) The dorsal aspect of the mid/hindbrain. (2) The dorsal-medial portion of the differentiated cranial somites. (3) Expression along the entire length of the unsegmented paraxial mesoderm (segmental plate) (Fig. 9A). At HH11, expression of *cNkd1* was evident in the medial portion of all but the youngest somite (Fig. 9B). At HH12, all somites expressed *cNkd1*. Cryosections taken from an HH12 embryo revealed not only the dynamic nature of the expression pattern in the paraxial mesoderm but also expression in tissues that were not apparent in whole embryos (e.g. expression in the notochord at the level of caudal segmental plate). Expression of *cNkd1* was detected throughout the segmental plate (Fig. 9G) and there was a significant up-regulation in the

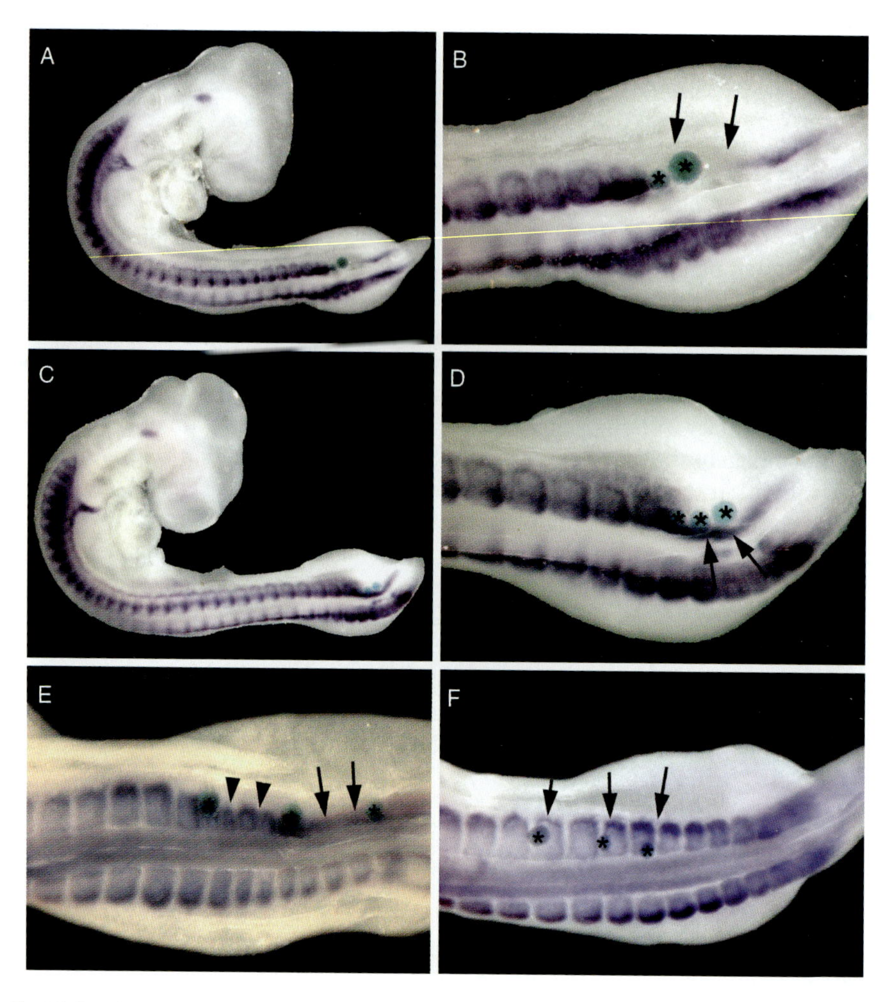

Fig. 8 A–F The Wnt antagonist Sfrp2 inhibits somite formation. **A** Whole mount in situ hybridisation for *Paraxis* of a 3-day-old chicken embryo after implantation of Sfrp2-soaked beads at the level of caudal segmental plate. **B** Higher magnification of the area of interest with the implanted beads shown in **A**. On the operation side at the level of beads, no somites and no *Paraxis* expression (*arrows*) are found (*asterisks*: beads). **C** Whole mount in situ hybridisation for *Paraxis* of a 3-day-old chicken embryo after implantation of Sfrp2-soaked beads at the level of caudal segmental plate. **D** Higher magnification of the area of interest with the implanted beads shown in **C**. On the operation side at the level of beads (*asterisks*: beads) only a very small band of *Paraxis* expression is found in direct contact to the neural tube (*arrows*). **E** Whole mount in situ hybridisation for *Pax3* of a 3-day-old chicken embryo after implantation of Sfrp2-soaked beads at the level of cranial and at the level of caudal segmental plate. On the operation side at the level of cranial beads (*asterisks*: beads), smaller *Pax3*-positive somites are found (*arrowheads*); at the level of caudal beads, no somites are evident, but *Pax3* expression is found (*arrows*). **F** Whole mount in situ hybridisation for *Pax3* of a 3-day-old chicken embryo after implantation of control beads (*asterisks*: beads). No effect on somite formation is visible (*arrows*)

cranial segmental plate where the tissue is thought to be undergoing a mesenchymal-to-epithelial transition (Fig. 9F). Expression of *cNkd1* was detected in the posterior notochord (at the caudal segmental plate level that was down-regulated at the level of epithelial somites) and posterior neural tube (Fig. 9G). Transcripts in both these tissues were down-regulated and not present in anterior regions (Fig. 9D–F). *cNkd1* was expressed throughout the epithelial somite but was concentrated at the dorsal medial aspect (Fig. 9E). In more mature somites, expression of *cNkd1* was found in a group of cells located at the dorsal medial aspect but also under the epithelially organised dermomyotome (Fig. 9D). Strong expression was detected in all somites at HH17 (Fig. 9H). Transverse sections through the embryo shown in Fig. 9H revealed that young somites maintained expression in the dorsal medial region (Fig. 9I), but that expression was thereafter up-regulated in more lateral regions immediately under the dermomyotome (Fig. 9J). However, expression was not found towards the hypaxial domain. Weak sclerotomal expression was additionally detected adjacent to the neural tube (Fig. 9J). We also detected expression of *cNkd1* during limb outgrowth. Initially, expression was confined to the dorsal sub-ectodermal mesenchyme (Fig. 9I). However by HH20, *cNkd1* was also expressed in the ventral equivalent (Fig. 9K–M'). In this stage, expression of *cNkd1* is re-activated in the neural tube, but unlike previous stages the expression of *cNkd1* is confined to the dorsal aspect.

3.2.2
Regulation of *cNkd1* Expression

Having established a comprehensive expression profile for *cNkd1* and shown that one of the major sites of expression was in the paraxial mesoderm, we subsequently carried out a series of surgical manipulations in order to identify the tissues that regulate the expression of the gene in the developing somites.

3.2.2.1
Medial Barrier Implantation

We implanted an impermeable barrier between the segmental plate and axial organs in order to assess the role of signals originating from the neural tube and notochord on the expression of *cNkd1*. This manipulation resulted in a down-regulation of *cNkd1* expression (Fig. 10A, B) in the paraxial mesoderm.

3.2.2.2
Neural Tube Removal

Since there was considerable down-regulation of *cNkd1* in the paraxial mesoderm following medial structure isolation, we further investigated the individual influences of the neural tube and notochord. In the first instance we determined the role of the neural tube on the expression of *cNkd1* by removing this structure at the level of the segmental plate. This procedure resulted in somites fused over the dorsal midline that failed to express *cNkd1* (Fig. 10C–D).

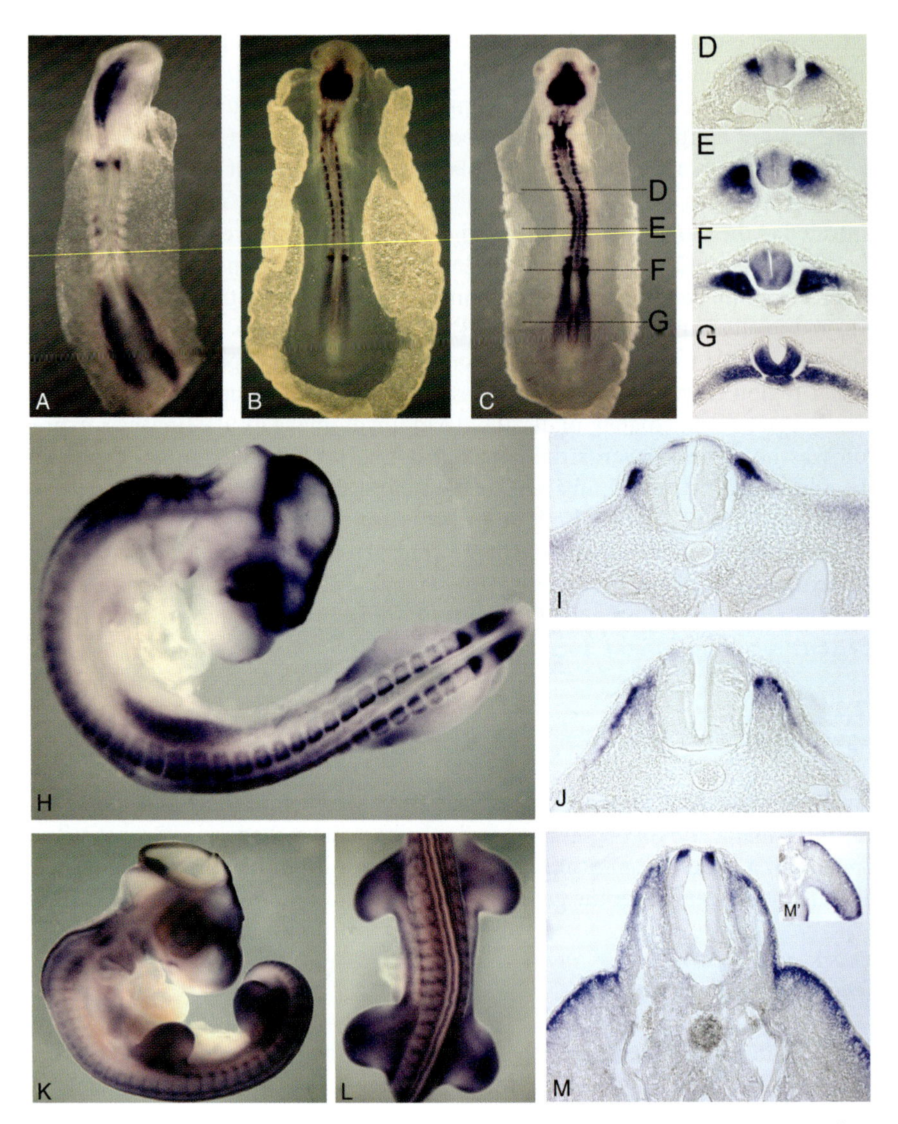

Fig. 9 A–M In situ hybridisation for *cNkd1* in HH stage 8–20 chick embryos. **A** Stage HH8 shows strong expression of *cNkd1* in the mid-and hindbrain region, the medial part of the most cranial somites and the segmental plate. **B** Stage HH11 shows strong expression of *cNkd1* in the head, in the medial portion of the somites and in the segmental plate. In the segmental plate itself the strongest expression is found in the most cranial part. **C** *cNkd1* expression at HH stage 12. All somites express the gene. Here differences between the somites become more evident. Differentiated somites have a strong expression in the medial part, while epithelial somites have an expression of *cNkd1* in the whole epithelium, with stronger expression in the dorsomedial part of the somite. Strong expression is also found in the segmental plate with a decreasing gradient from cranial to caudal. **D–G** Transverse sections of the embryo shown in **C**. **D** Section at the level of the differentiated somite. Expression of *cNkd1* is found in the medial part of the ingressing myotome and the adjacent dorsomedial part of the sclerotome.

3.2.2.3
Notochord Removal

In contrast to the down-regulation of *cNkd1* expression observed after neural tube ablation, we found that the removal of the notochord resulted in an up-regulation in gene expression in the medial and ventral part of the somites as well as the dorsal neural tube (Fig. 10E–F).

3.2.3
Regulation of *cNkd1* by Members of the Wnt Family

Previous work has shown that members of the Wnt family of signalling molecules can regulate the expression of *Nkd* (Zeng et al. 2000). Our tissue manipulation studies have shown that signals originating from the neural tube are required to maintain the expression of *cNkd1* in the paraxial mesoderm. In these experiments, we implanted cells expressing various Wnt proteins either medial or lateral to the segmental plate to determine their effect on *cNkd1* expression. We found that implantation of *Wnt1*-, *Wnt3a*- or *Wnt4*-expressing cells medial or lateral to the segmental plate (Fig. 11A–L) resulted in an up-regulation of expression of *cNkd1* in the medial part of the newly formed somites (Fig. 11A–L).

In contrast to the up-regulation of *cNkd1* expression induced by Wnt proteins that are expressed in the neural tube, we found that implantation of *Wnt6*-expressing cells medial to the segmental plate (Fig. 11M–N) resulted in a down-regulation of *cNkd1* expression in the somites. The lateral implantation of Wnt6 cells had no effect on the *cNkd1* expression in the somites (Fig. 11O–P). Implantation of *Wnt11*-expressing cells medial or lateral to the segmental plate resulted in a down-regulation of *cNkd1* expression in the newly formed somites (Fig. 11Q–T). We have shown therefore that Wnt1, Wnt3a and Wnt4 are able to up-regulate the expression of *cNkd1* in the somites,

◄

Fig. 9 A–M (continued) **E** Section at the level of the epithelial somite. Expression of *cNkd1* is found in the whole somite with the strongest expression in the dorsomedial part and the weakest expression in the ventrolateral part. **F** Section at the level of the cranial segmental plate. Expression is found in the partially epithelialised cranial segmental plate. **G** Section at the level of the caudal segmental plate. Expression of *cNkd1* is found in the segmental plate and at weaker extent the intermediate and lateral plate mesoderm. **H** At stage HH17 expression of *cNkd1* is found in the head, somites, segmental plate and in the limb buds. **I** Section of the embryo shown in **H** at the level of the cranial part of the caudal limb bud. Expression of *cNkd1* is found in the ingressing myotome and adjacent sclerotome. A weaker expression is also found in the most dorsal part of the neural tube and in the dorsal sub-ectodermal mesenchyme lateral to the dermomyotome. **J** Section of the embryo shown in **H** at the interlimb level. Expression of *cNkd1* is found in the myotome beneath the dermatome now extending more lateral and in the most dorsomedial part of the sclerotome. **K, L** Stage HH20 shows expression of *cNkd1* in the head, somites, limb buds and in the dorsal neural tube. **M, MI'** Section of the embryo shown in **K** reveals expression of *cNkd1* in the sub-ectodermal mesenchyme of the somite and limb bud and expression in the dorsal part of the neural tube

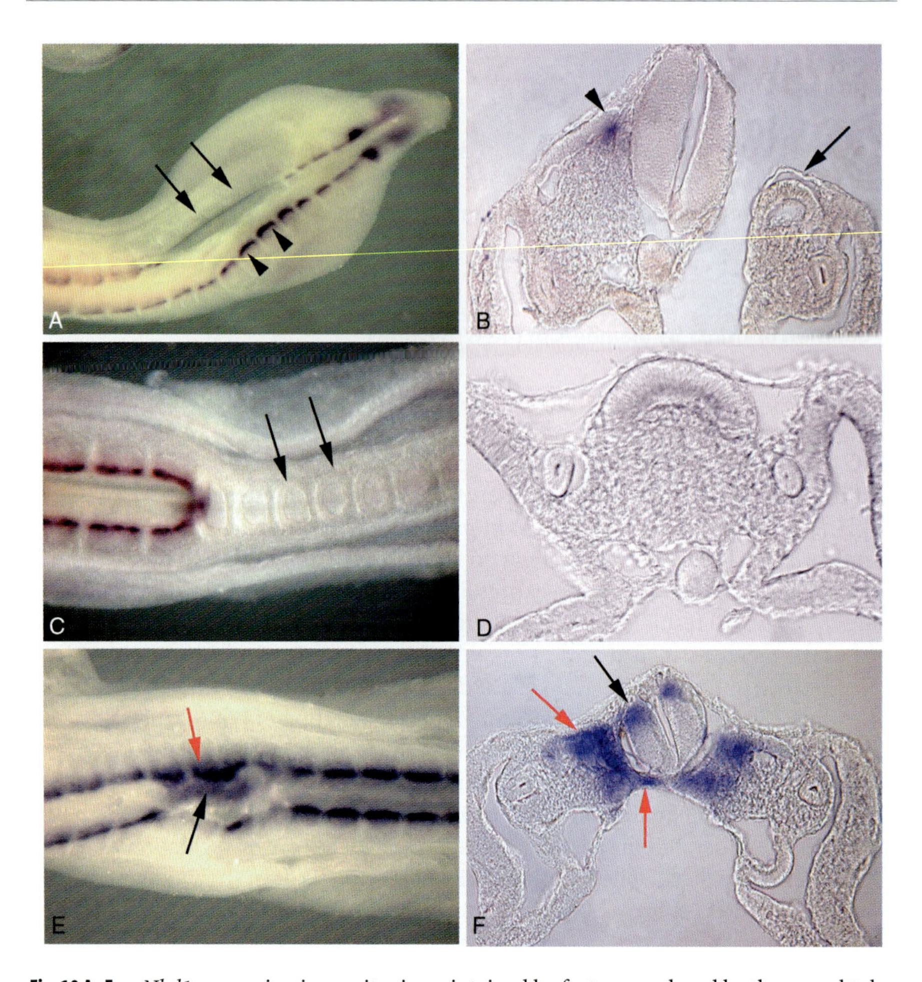

Fig. 10 A–F *cNkd1* expression in somites is maintained by factors produced by the neural tube. **A** Whole mount in situ hybridisation for *cNkd1* of a 3-day-old chick embryo after implantation of an impermeable barrier between neural tube/notochord and the segmental plate and following re-incubation for 20 h. On the operation side, no expression of *cNkd1* is found (*arrows*) compared to the control side, where there is strong *cNkd1* expression in the medial part of the somites (*arrowheads*). **B** Transverse section of the embryo shown in **A** shows no expression in the somite separated from the neural tube (*arrow*) compared to the somite of the control side with *cNkd1* expression in the medial part of the somite (*arrowhead*). **C** Whole mount in situ hybridisation for *cNkd1* of a 3-day-old chick embryo after removal of the neural tube at the level of segmental plate and following re-incubation for 20 h. There is no expression of *cNkd1* in the newly formed fused somites (*arrows*). **D** Transverse section of the embryo shown in **C** shows a fused single somite without *cNkd1* expression. **E** Whole mount in situ hybridisation for *cNkd1* of a 3-day-old chick embryo after removal of the notochord at the level of the segmental plate and following re-incubation for 20 h. At the operation level there is a stronger *cNkd1* expression in the newly formed somites (*red arrow*) and in the neural tube (*black arrow*). **F** Transverse section of the embryo shown in **E** reveals a stronger expression of *cNkd1* in the somite (*red arrows*) and in the dorsal neural tube (*black arrow*) at the operation level

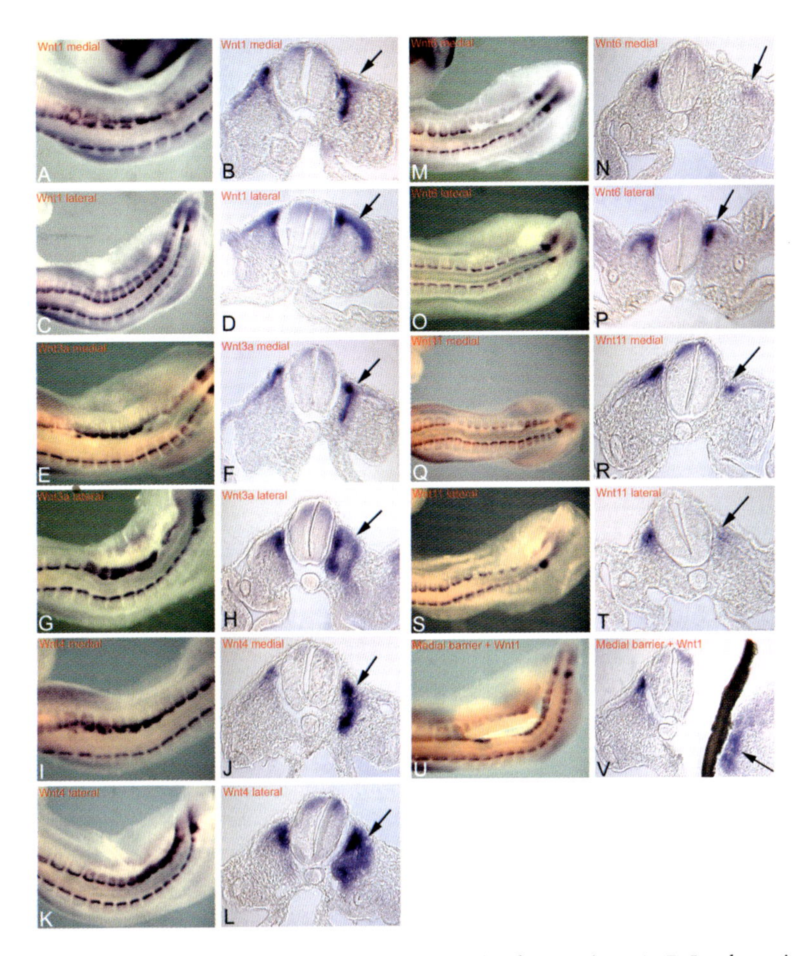

Fig. 11 A–V Wnt proteins control *cNkd1* expression in the somites. **A, B** Implantation of *Wnt1*-expressing cells medial to the segmental plate leads to an up-regulation of *cNkd1* in the somites at the operation level (*arrow*). **C, D** Implantation of *Wnt1*-expressing cells lateral to the segmental plate leads to an ectopic expression of *cNkd1* in the lateral part of the somites at the operation level (*arrow*). **E, F** Implantation of *Wnt3a*-expressing cells medial to the segmental plate leads to an up-regulation of *cNkd1* in the somites at the operation level (*arrow*). **G, H** Implantation of *Wnt3a*-expressing cells lateral to the segmental plate leads to an ectopic expression of *cNkd1* in the lateral part of the somites at the operation level (*arrow*). **I, J** Implantation of *Wnt4*-expressing cells medial to the segmental plate leads to an up-regulation of *cNkd1* in the somites at the operation level (*arrow*). **K, L** Implantation of *Wnt4*-expressing cells lateral to the segmental plate leads to an ectopic expression of *cNkd1* in the lateral part of the somites at the operation level (*arrow*). **M, N** Implantation of *Wnt6*-expressing cells medial to the segmental plate leads to a down-regulation of *cNkd1* in the somites at the operation level (*arrow*). **O, P** Implantation of *Wnt6*-expressing cells lateral to the segmental plate had no effect of the expression of *cNkd1* in the somites. **Q–T** Implantation of *Wnt11*-expressing cells medial or lateral to the segmental plate leads to a down-regulation of *cNkd1* in the somites at the operation level (*arrow*). **U, V** After implantation of an impermeable barrier and *Wnt1*-expressing cells at the level of the segmental plate, expression of *cNkd1* is maintained through Wnt1 in the separated somites at the operation level (*arrow*)

whereas Wnt11, and to a lesser degree Wnt6, both cause a down-regulation of its expression. We finally asked whether Wnts alone could maintain *cNkd1* expression in the paraxial mesoderm. To this end, we implanted an impermeable barrier between axial structures and the segmental plate, but in addition placed Wnt1- or Wnt3a-expressing cells beneath the ectoderm overlying the segmental plate. In contrast to the complete down-regulation following medial barrier implantation, we found that Wnt1 (Fig. 11U, V) and Wnt3a (data not shown) were able to maintain *cNkd1* expression even after the influence of the neural tube had been removed.

Implantation of control (LacZ cells) medial or lateral to the segmental plate had no effect on the expression of *cNkd1* (data not shown).

3.3 Wnt6 and Neural Crest

3.3.1 Wnt6 Induces Neural Crest Production and Wnt1 Inhibits Neural Crest Induction

The ability of Wnt1, Wnt3a, Wnt4, Wnt6 and Wnt11 to induce neural crest in the chick embryo was examined by implanting *Wnt*-expressing cells into the open caudal neural plate at HH10–12. At this axial level, markers of neural crest development such as *Slug*, *Sox10*, *FoxD3* and *RhoB* are not expressed in the dorsal neural folds (Cheung and Briscoe 2003; Liu and Jessell 1998). We monitored the expression of the neural crest markers *Slug*, *RhoB*, *Sox10*, *FoxD3*, and HNK-1 8–12 h after implantation. Using this assay, we found that cells expressing *Wnt3a*, *Wnt4* or *Wnt11* had no effect on the induction of neural crest (data not shown). In contrast, *Wnt6*-expressing cells induced ectopic expression of early neural crest markers (Fig. 12B, D, F). These markers were ectopically expressed in caudal regions of the embryo adjacent to the implanted cells. In addition, we found that *Wnt1*-expressing cells (Fig. 12H) and LacZ control cells (Fig. 12A, C, E, G) did not induce the expression of neural crest markers. The effect of Wnt6 and Wnt1 on neural crest induction was quantified in vitro by plating identical portions of neural tube from the level of the segmental plate onto *Wnt*-expressing cells and counting the number of cells expressing the neural crest marker HNK-1 in eight individual cultures for each cell type after 20–24 h. In Wnt6 cultures, an average of 230±9 cells were produced in each culture, compared to 37±1.2 in each LacZ control culture, equating to a sevenfold increase in neural crest cell production (Fig. 12I). Surprisingly, culture of neural tubes on *Wnt1*-expressing cells resulted in a complete inhibition of crest induction compared to the LacZ controls, with no HNK-1-positive cells in any of the experiments (Fig. 12I). This latter result was confirmed in vivo by analysing *Sox10* expression in embryos 40 h after implantation of *Wnt1*-expressing cells. There was a significant reduction in the amount of neural crest generated in the region where cells were implanted (Fig. 12J, K). These results are significant since, first, they highlight a novel role for Wnt6 as a neural crest inducer protein, and second, we show that Wnt1, a previous candidate for the neural crest inducer, actually inhibits the production of these cells.

3.3.2 Wnt6 Induces Neural Crest Production Through the Non-canonical Signalling Pathway

We subsequently examined the intracellular signalling cascade utilised by Wnt6 during neural crest induction, as previous studies have provided evidence that it can signal through both the canonical and non-canonical pathway (Itaeranta et al. 2002; Buttitta et al. 2003). First, we examined whether Wnt6 acts through the canonical pathway by examining the localisation of activated β-catenin, as its nuclear localisation is predictive of activation of the canonical pathway. In Wnt6-implanted embryos, nuclear localisation of activated β-catenin was not found in the dorsal neural tube cells (Fig. 13B, C), suggesting that the canonical pathway was not active. In contrast, activated β-catenin was found in the nuclei of dorsal neural tube cells in Wnt1-implanted embryos (Fig. 13D, E), implying that the canonical pathway was active in these cells. The nuclei of these cells appeared condensed, suggesting they are undergoing apoptosis. When the dorsal neural tube of Wnt6- and Wnt1-implanted embryos was examined using TUNEL labelling, however, there was no difference in the pattern of apoptosis seen in either treatment (Fig. 13F, G), with very few apoptotic cells in both cases. To determine whether the non-canonical signalling pathway was activated by Wnt6, we examined the expression of two downstream components of this pathway, *RhoA* (Kim and Han 2005) and Jun N-terminal kinase (JNK) (Li et al. 1999). *RhoA* and activated (phosphorylated) JNK expression was found in dorsal neural tube cells when *Wnt6*-expressing cells had been implanted (Fig. 13I, K), indicating that the non-canonical pathway was active. In contrast, implantation of control cells (Fig. 13) or *Wnt1*-expressing cells (Fig. 13J) did not result in the expression of either molecule in cells of the dorsal neural tube. This is a clear indication that Wnt6 and Wnt1 are working through separate pathways, resulting in opposing effects on neural crest induction in the amniote.

It has previously been suggested that nuclear localisation of activated β-catenin can be seen in the dorsal neural tube during neural crest induction (i.e. in the open neural plate) at stage 10 (Garcia-Castro et al. 2002). We examined unoperated embryos between stages 9 and 12 for activated β-catenin localisation along the whole antero-posterior axis of the neural tube (data not shown) but were only able to identify nuclear staining in the most cranial regions.

3.3.3 Activation of the Non-canonical Pathway Through Dsh Induces Neural Crest

To prove that the non-canonical pathway induces neural crest and the canonical cascade inhibits this process, we specifically inhibited each pathway with Dsh deletion mutant constructs. Dsh is a component of both the canonical and non-canonical

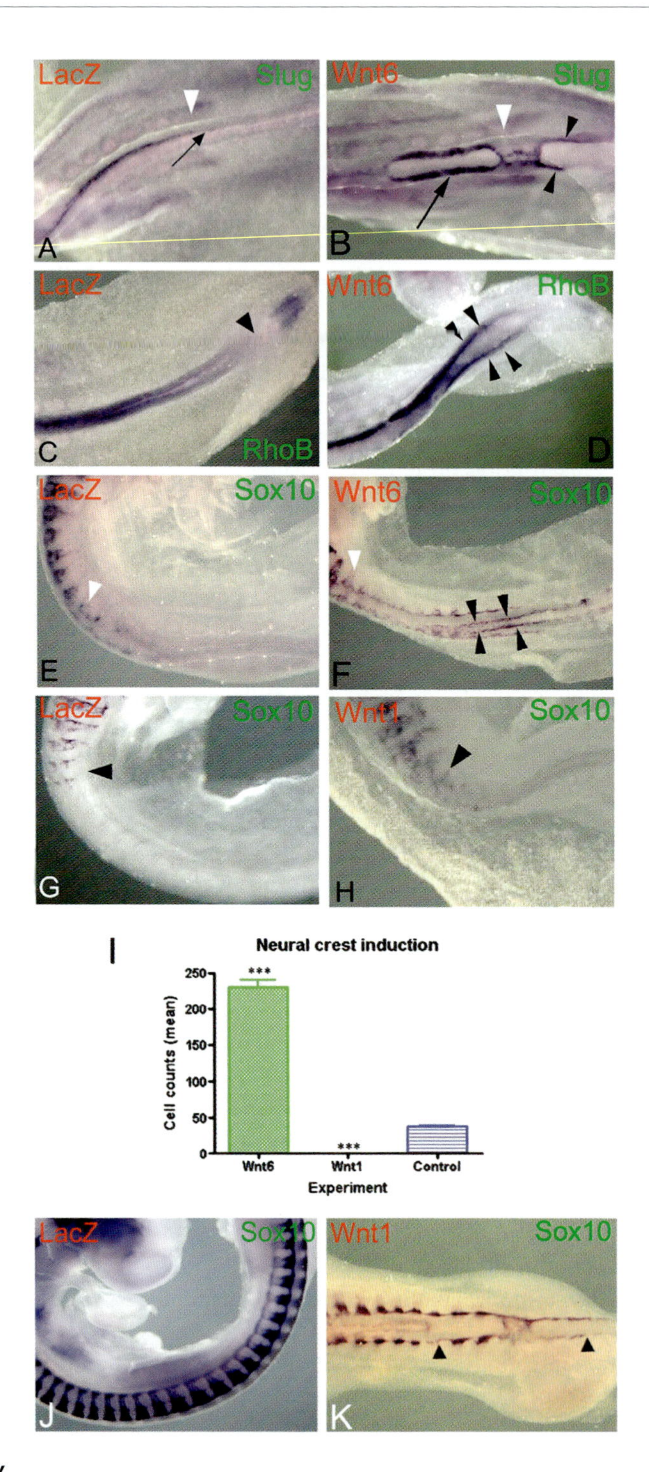
LacZ
Slug
A
Wnt6
Slug
B
LacZ
RhoB
C
Wnt6
RhoB
D
LacZ
Sox10
E
Wnt6
Sox10
F
LacZ
Sox10
G
Wnt1
Sox10
H
I
Neural crest induction
***
250
200
150
100
50
0
Cell counts (mean)
***
Wnt6
Wnt1
Control
Experiment
LacZ
Sox10
J
Wnt1
Sox10
K

Fig. 12 A–K

signalling pathway and activation of each pathway is controlled by specific domains in the Dsh protein; the non-canonical pathway requires the DEP domain whereas the canonical pathway requires the DIX domain (Axelrod et al. 1998; Li et al. 1999). Deletion of these specific domains blocked signalling through each specific pathway (Krylova et al. 2000; Rosso et al. 2005). These inhibiting constructs were individually electroporated into one side of the chick caudal open neural plate between HH10 and HH12. Effects on neural crest were assayed by analysis of Slug and Sox10 expression. Electroporation of the Δ-DIX-Dvl construct, which prevents canonical signalling allowing the non-canonical pathway to predominate, resulted in an increased number of premigratory Slug-positive cells in the caudal neural tube at stage 15 (Fig. 14A). Sections demonstrate that while crest is induced, it does not migrate prematurely (Fig. 14B). In addition, migratory Sox10-positive cells at the dorsal edge of the neural tube and in the periphery were also increased at stage 19 compared to the contra-lateral control (Fig. 14C, D), similar to the implantation of Wnt6 cells (Fig. 12B, F). The Δ-DEP-Dvl construct, which prevents signalling through the non-canonical pathway, allowing the canonical pathway to predominate, resulted in a significant reduction in neural crest production (Fig. 14E, F) and mirrored the effect of Wnt1 over-expression (Fig. 12H–K).

3.3.4 The Canonical Wnt Pathway Inhibits Neural Crest Induction

Our results suggest that the non-canonical and canonical pathways may act antagonistically in the context of neural crest induction. To test this hypothesis

Fig. 12 A–K Neural crest induction after implantation of Wnt6- and Wnt1-expressing cells. Expression of neural crest markers, assayed by in situ hybridisation, **A–H** 8–12 h after implantation of cells and **J, K** 40 h after implantation of cells. **A** *Slug* expression in the control dorsal neural tube (*black arrow*) extends to the level of the last formed somite (indicated by *white arrowhead*). **B** Wnt6 induces higher levels (*black arrow*) and ectopic (*black arrowheads*) expression of Slug caudal to the last somite (*white arrowhead*). **C** *RhoB* expression is absent in the caudal neural tube (*arrowhead*) but present in the tail bud. **D** Wnt6 increases *RhoB* expression in the caudal neural tube (*arrowheads*). **E** *Sox10* is expressed in rostral migrating neural crest in control embryos. *White arrowhead* indicates caudal limit of migration. **F** Wnt6 induced ectopic *Sox10* expression caudally compared to control (*black arrowheads*). The *white arrowhead* is the equivalent somite level as in **E**. **G, H** Wnt1 does not induce neural crest. The *arrowheads* indicate the equivalent level in neural tube. $n = 20/20$ for all markers. All at ×8 magnification. **I** In vitro analysis of neural crest induction by Wnt1 and Wnt6. Wnt6 induced a sevenfold increase in the numbers of neural crest cells compared to control. Wnt1 produced an inhibition of neural crest production compared to controls. Analysis by two-tailed *t*-test indicates that both Wnt6 and Wnt1 are highly significantly different from the controls at the 0.0001% level (indicated by ***). **J, K** Analysis of *Sox10* expression in embryos 40 h after implantation of Wnt1-expressing cells confirms that Wnt1 inhibits neural crest production (*black arrows*). $n = 7/7$

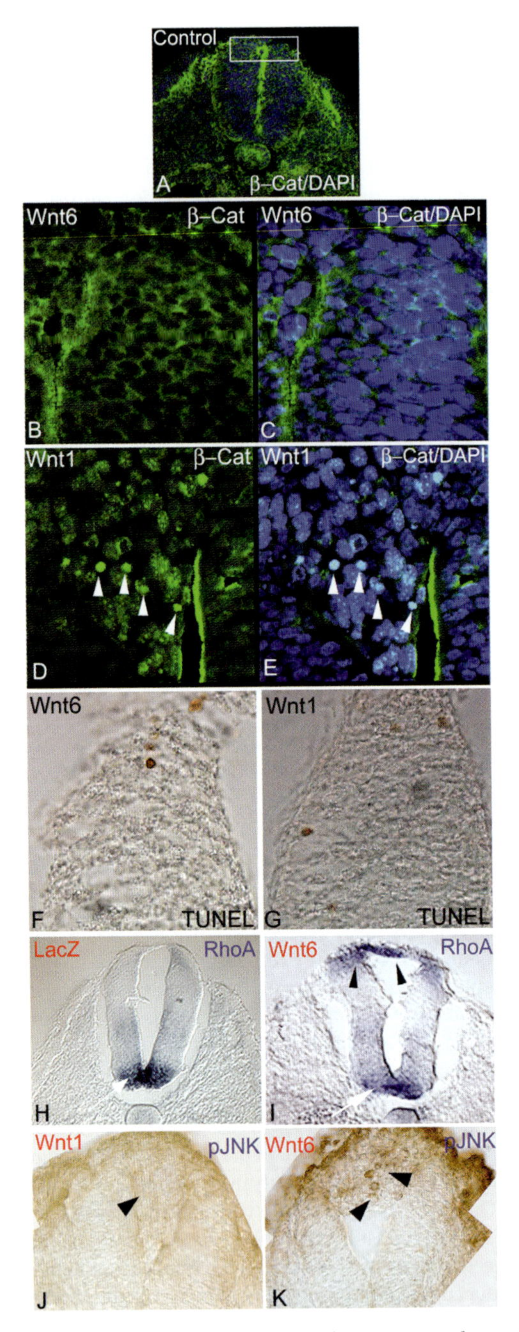

Fig. 13 A–K Wnt6 signals through the non-canonical pathway in neural crest induction. Embryos are at stages HH16–18. **A** The control section is labelled with activated β-catenin antibody (*green*) and DAPI (*blue*). The *box* indicates area shown at higher magnification in **B–E. B, C** Wnt6 cells do not induce activated β-catenin translocation to the nucleus in dorsal neural tube cells.

we either amplified or attenuated canonical signalling and examined the outcome of such action on neural crest production. Activation of the canonical pathway by over-expression of a constitutively active β-catenin construct (Baker et al. 1999) in the neural tube resulted in a reduction in neural crest formation (Fig. 14G, H). We performed the complementary experiment in which we inhibited Wnt1 activity at two different points in the signalling pathway. Over-expressing a dominant-negative form of TCF-4 (Tetsu and McCormick 1999) resulted in both increased and ectopic induction of neural crest (Fig. 14 I–L) as shown with both *Slug* and *Sox10* expression. Second, over-expression of *Naked cuticle* (*Nkd*), an inhibitor of canonical Wnts (Rousset et al. 2001), resulted in a similar but stronger phenotype, with increased and ectopic neural crest production in the electroporated side (Fig. 14M, N).

3.3.5 Specific Inhibition of Wnt6 Reduces Neural Crest Production

These experiments indicate that Wnt6 can induce neural crest; however, they do not definitively prove that Wnt6 is required for this function in the embryo. To demonstrate that Wnt6 is necessary for neural crest induction, we developed two RCAS siRNA constructs (T1 and T2) to reduce endogenous levels of Wnt6 protein. siRNA containing RCAS viral particles were injected onto the ectoderm adjacent to the developing neural tube at HH7–11 and allowed to develop for 2–3 days. Viral spread was mainly restricted to the ectoderm, with a small amount in the underlying tissue (assessed by GAG labelling, Fig. 15B, inset i). These reagents caused a considerable decrease in the level of Wnt6 protein in the ectoderm adjacent to and overlying the dorsal neural tube (compare Fig. 15A and its inset panel with Fig. 15B and inset panel ii), comparable to levels achieved previously (Das et al. 2006). The reduction in ectodermal Wnt6 corresponded to the site of siRNA viral construct infection (Fig. 15B, insets i and ii). Construct T2 had greater efficacy than T1 and was selected for use in further studies. Analysis of the neural crest markers *FoxD3* and HNK-1 showed a marked reduction in neural crest at the dorsal edge of the neural tube and migrating into the periphery (Fig. 15B, D, F, n=4/5) compared to controls (Fig. 15A, C, E). These results indicate that neural crest induction has been inhibited. At later stages (HH22) there were fewer neural crest cells in the adjacent somites, indicated by *Sox10* expression (data not shown). *Wnt1* and *Wnt3a* expression were not affected (data not shown), indicating the specificity of the siRNA

Fig. 13 A–K (continued) **D**, **E** Wnt1 induces translocation of β-catenin to the nucleus in many dorsal neural tube cells. *Arrowheads* indicate numerous labelled nuclei. **F** Wnt6-implanted and **G** Wnt1-implanted embryos do not show differing levels of TUNEL labelling. **H** The control embryo demonstrates *RhoA* is expressed ventrally (*arrow*). **I** Local induction of *RhoA* in the dorsal neural tube by Wnt6 (*arrowheads*). **J** In Wnt1 implanted embryos, there is no pJNK labelling. **K** Wnt6-induced cells in the neural tube are pJNK positive (*arrowheads*). n=7/7 for each experiment. **A** ×10 magnification, **B–E** ×40 magnification, **F**, **G** ×64 magnification, **H–K** ×20 magnification

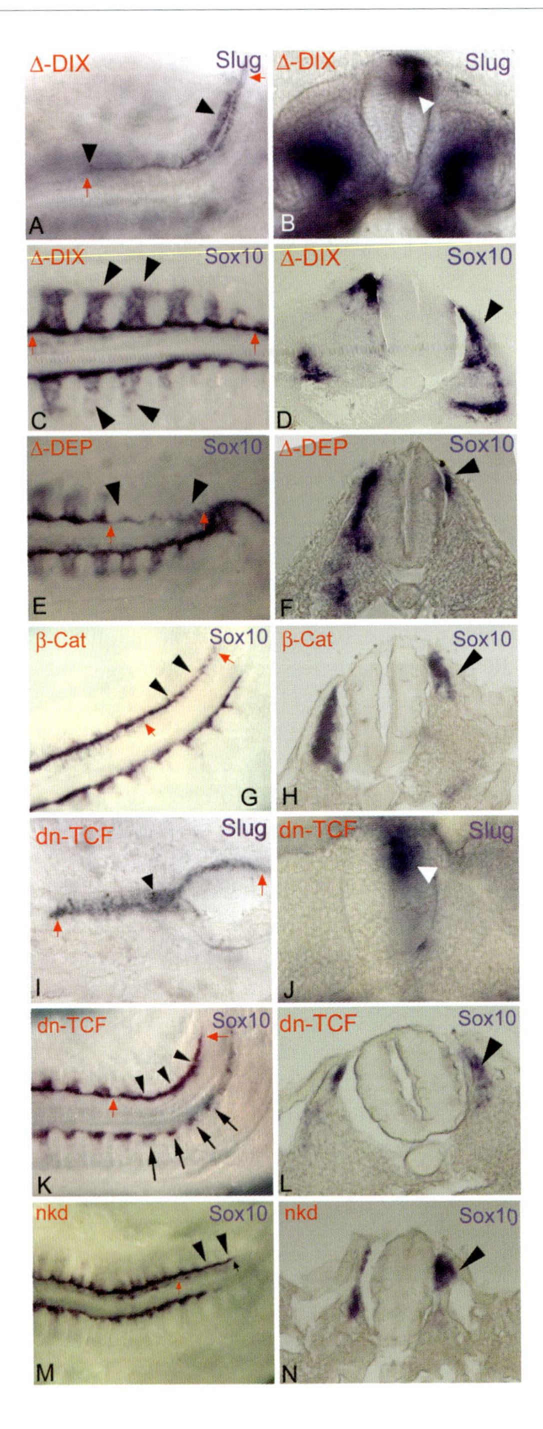
Δ-DIX
Slug
A
Δ-DIX
Slug
B
Δ-DIX
Sox10
C
Δ-DIX
Sox10
D
Δ-DEP
Sox10
E
Δ-DEP
Sox10
F
β-Cat
Sox10
G
β-Cat
Sox10
H
dn-TCF
Slug
I
dn-TCF
Slug
J
dn-TCF
Sox10
K
dn-TCF
Sox10
L
nkd
Sox10
M
nkd
Sox10
N

Fig. 14 A–N

construct. Thus, specific inhibition of *Wnt6* expression demonstrates that it is required for neural crest induction in the chick embryo.

4 Discussion

4.1 Wnt6 and Somitogenesis

The processes regulating the development and maturation of the somites are being slowly unravelled. Whereas the segmentation of the paraxial mesoderm is driven by intrinsic properties (Cooke and Zeeman 1976; Pourquie 2000), the organisation of the same tissue and subsequent differentiation is regulated by extrinsic factors originating from adjacent tissues (Brand-Saberi et al. 1996; Rong et al. 1992). There is a clear distinction between segmentation and somitogenesis. Segmentation represents formation of repetitive portions of the paraxial mesoderm whereas somitogenesis is the epithelialisation of the segmentally organised paraxial mesoderm. Deletion of numerous genes has shown that somitogenesis is not a prerequisite for the differentiation of somitic derivatives (Burgess et al. 1996; Hrabe de Angelis et al. 1997; Johnson et al. 2001). However, although the Paraxis−/− mouse (which fails to form somites) still generates muscle and axial skeleton these tissues are disorganised (fused vertebrae, fused ribs and patterning defects of axial musculature; Burgess et al. 1996). The process of somitogenesis seems therefore to regulate the precise organisation and growth of tissues, allowing them to eventually differentiate in an orderly manner.

◄

Fig. 14 A–N Inhibition of the non-canonical or canonical pathway during neural crest induction. Unilateral electroporation of half of the neural tube was assayed by either *Slug* expression at st14/15 or *Sox10* expression at HH18/19. *Red arrows* indicate the region of the neural tube electroporated. **A, B** Inhibition of the canonical pathway by Δ-DIX-Dvl increased the numbers of Slug-expressing neural crest cells in the dorsal neural tube and periphery (*black arrowheads*). **C, D** Inhibition of the canonical pathway by Δ-DIX-Dvl increased the numbers of *Sox10*-expressing neural crest cells in the dorsal neural tube and periphery (*black arrowheads*). **E, F** Inhibition of the non-canonical pathway by Δ-DEP-Dvl inhibited neural crest production in the dorsal neural tube (*arrowheads*), leading to a subsequent absence of crest in the periphery. **G, H** Over-expression of *β-catenin* reduced *Sox10* expression in the electroporated side (*arrowheads*) compared to the contra-lateral (control) side. **I, J** Inhibition of the canonical pathway with *dn-TCF* increases *Slug* expression in the neural tube (*arrowhead*). **K, L** Inhibition of the canonical pathway with *dn-TCF* increases *Sox10* expression in the neural tube (*arrowheads*), resulting in the loss of the normal segmented pattern seen on the control side (*arrows*). **M, N** Inhibition of the canonical pathway with *nkd* increased the normal expression of *Sox10* (*red arrows*), and also induced ectopic *Sox10* expression in the caudal neural tube (*arrowheads*). *n*=7/7 for all experiments. **A, C, E, G, I, K, M** ×10 magnification. **B, D, F, H, J, L, N** ×20 magnification

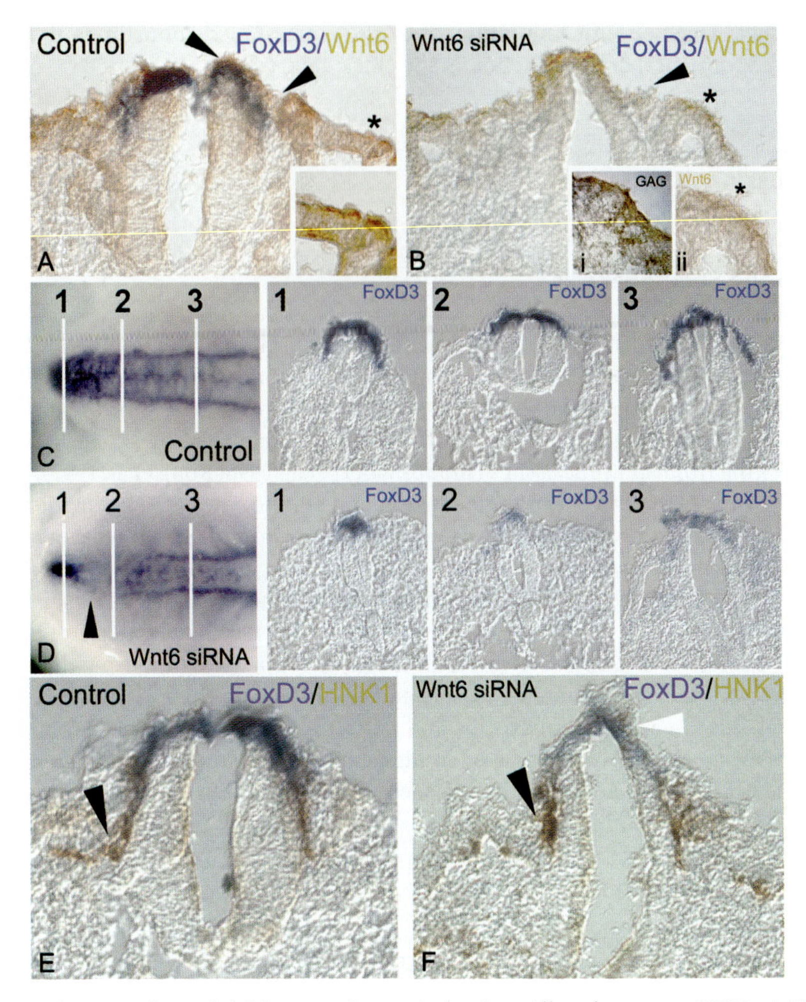

Fig. 15 A–F Wnt6 siRNA inhibits neural crest induction. All embryos are HH18. **A** Wnt6 protein expression in control embryos. Wnt6 is strongly expressed in the ectoderm (*black arrowheads*). The *inset panel* shows the region labelled with an *asterisk* at higher magnification. *FoxD3* is expressed in the dorsal neural tube. **B** Wnt6 expression in Wnt6 siRNA-treated embryos. Wnt6 levels are much reduced (*arrowhead*). Inset panel *i* shows GAG labelling, which demonstrates the extent of viral spread, mainly in the ectoderm with some in underlying tissue. Inset panel *ii* shows region labelled with an *asterisk* in the main image. *FoxD3* expression is absent in the dorsal neural tube and Wnt6 levels in the ectoderm are much reduced. **C** *FoxD3* expression in neural crest. *Lines* indicate the level of sections displayed in the adjacent panels. *FoxD3* is expressed in neural crest in the dorsal neural tube and in the periphery. **D** *FoxD3* expression in embryo treated with Wnt6 siRNA. Again, *lines* indicate levels of adjacent sections. Expression of *FoxD3* is significantly reduced in regions of the neural tube (*arrowhead* and *panel 2*, *3*) and in the periphery. **E** HNK-1 and *FoxD3* expression in the control embryo. The *arrowhead* indicates migrating crest. **F** HNK-1 and *FoxD3* expression in Wnt6 siRNA-treated embryo. *FoxD3* expression is significantly reduced in the neural tube (*white arrowhead*). n=4/5. **A**, **B**, **E**, **F**, ×20 magnification. **C**, **D** ×10 magnification

This study has identified Wnt6 originating from the dorsal ectoderm as a possible candidate that can regulate the process of somite formation. The role of ectodermal Wnt6 identified in this study seems to be similar to the role played by Paraxis, i.e. they both promote epithelialisation. We showed that removal of ectoderm leads to the loss of *Paraxis* and that Wnt6 rescues both *Paraxis* expression and epithelialisation of segmental plate. *Paraxis* expression is initiated in the segmental plate. Therefore, we suggest that Wnt6 signalling could either initiate or stabilise the expression of *Paraxis*, facilitating epithelialisation. The above experiments have shown that Wnt6 can influence the epithelialisation of the paraxial mesoderm. Since the application of the Wnt antagonist Sfrp2 interferes with somite epithelialisation, this demonstrates that Wnts play an active role in this process. In the cranial segmental plate, a region marked by the expression of *Paraxis*, SFRP2 did not totally prevent epithelialisation but led to the formation of small somites. Application of SFRP2 to the caudal, non-*Paraxis* expressing region of the segmental plate, however, completely inhibited epithelialisation. The size of the *Pax3* expression domain was reduced, but levels were normal in the small Sfrp2-induced somites. When somite formation was totally inhibited, *Pax3* expression levels were still found similar to the expression of this gene following genetic deletion of Paraxis (Burgess et al. 1995). Furthermore, we show that Wnt6 specifically regulates the expression of genes involved in the epithelialisation process. The role of promoting epithelialisation performed by Wnt6 seems to be unique amongst this family. We have previously investigated the role of Wnt1, Wnt3a and Wnt4 and found that they were involved in regulating myogenesis (Wagner et al. 2000).

In this study we showed that transplantation of cells expressing *Wnt6* beneath the ectoderm overlying the caudal segmental plate leads to an up-regulation of markers associated with epithelial somites and reduced levels of *Pax1*, a gene associated with the mesenchymal compartment of the somites. Furthermore, the expression of *MyoD* was greatly reduced as shown in the limb (Fig. 5A and C). However, there was no evidence of cell loss as the somites appeared free of dying cells.

These observations suggest *Wnt6* over-expression leads to cells becoming locked in an epithelial state for an extended period. Cells are eventually released from the epithelial state to continue their differentiation programme, as demonstrated by the presence of axial skeleton (data not shown) and muscle in old embryos.

Application of Wnt6-producing cells led to a reduction in the amount of muscle in the limbs, and we were able to show a pronounced decrease in the expression of markers associated with migrating muscle precursors. All limb muscles originate from the somites (Christ et al. 1977). Somitic cells migrate as un-differentiated precursors into the limb bud, where they execute their myogenic programme of development. Since the limb outgrowth programme is not affected by the implantation of Wnt6-producing cells, we assume that the expression of the migration-inducing agent *Scatter Factor* (Brand-Saberi et al. 1996; Dietrich et al. 1999; Scaal et al. 1999) is not altered but is nevertheless insufficient to induce limb muscle migration. These results suggest that there must also be a modulation of the epithelial organisation of the hypaxial lip in order to facilitate the migratory process.

Wnts can activate a number of signalling pathways (Dale 1998), each with a distinct consequence in terms of cellular development. The two most commonly used are the canonical and non-canonical Wnt pathways. The former involves the activation of a signalling process that results in translocation of β-catenin from the cytoplasm to the nucleus and leads to the initiation of gene transcription (Gumbiner 1995; Huber et al. 1996; Willert and Nusse 1998). This signalling cascade is deployed during tissue patterning. In contrast, the non-canonical (or in *Drosophila*, the planar cell polarity, PCP) pathway signals through small GTPases and activates c-Jun N-terminal kinases to control cell behaviour and arrangement (e.g. epithelialisation) rather than identity (Weston and Davis 2002). However, β-catenin does have a role in the non-canonical as well as the canonical pathway. E-cadherins are calcium-dependent transmembrane adhesion molecules found on epithelial cells and are linked to the cytoskeleton by β-catenin and form the adherens junction complex—a key feature of epithelial tissue. The enzyme glycogen synthase kinase 3 activates the proteasome-mediated degradation of free intracellular β-catenin (Peifer and Polakis 2000). Translocation of free β-catenin to the plasma membrane where it complexes with E-cadherin and actin filaments, however, not only prevents it from being degraded but also confers epithelial characteristics to the cell, without protein translation or gene transcription (Wiggan and Hamel 2002). Truncation of the β-catenin protein renders it unable to participate in the formation of adherens junctions and leads to a loss in the intercellular adhesiveness of cells, which has been implicated in the metastasis process (Oyama et al. 1994). Furthermore, the translocation of β-catenin to the nucleus suppresses the transcription of E-cadherin, again resulting in decreased cell adhesion (Jamora et al. 2003). In the case of the epithelialisation of the segmental plate, it could be that Wnt6 acts to stabilise adherens junctions possibly by translocating β-catenin to the plasma membrane and not the nucleus, where it may suppress the expression of E-cadherin (Jamora et al. 2003). But we cannot rule out that the non-canonical signalling pathway may also be involved in somite formation. Further studies have to show which pathway is used by Wnt6 to establish and maintain an epithelial pattern in the paraxial mesoderm.

Results from this and other studies establish the need for us to develop an epithelial somite (for the organisation of a differentiated derivative) and a requirement for the temporal regulation of this organisation. If cells are locked into an epithelial state then the cells cannot undergo their normal programme of differentiation as seen by the delayed onset of genes marking the differentiation of the sclerotome or myotome. We have provided conclusive evidence that Wnt6, which is expressed in the dorsal ectoderm, can substitute for the dorsal ectoderm in its ability to regulate the development of epithelial somites. We have also noted that there was a remnant of an epithelial somite at the dorsal medial aspect following ectoderm ablation, suggesting that factors others than Wnt6 can initiate epithelialisation. This action could be mediated through another member of the Wnt protein family since numerous members are expressed in this region (Cauthen et al. 2001). Since epithelial organisation was lost following medial barrier implantation, when the dorsal ectoderm was intact, this would suggest that a signal from medial structures could be responsible for the maintenance of *Wnt6* expression in the ectoderm or for the induction of the epithelialisation of the segmental plate.

During early somitogenesis each somite displays both mesenchymal and epithelial organisation. One means by which this could be achieved would be to limit the range over which Wnt6 diffuses. However, since the epithelial-to-mesenchymal transition seen in somites is at the medial-ventral portion, it is difficult to envisage how a gradient of Wnt6 could be generated from a relatively flat epithelium to solely target this region. We suggest that the medial-ventral portion of the somite escapes the epithelialising properties of Wnt6 through the action of antagonists. Furthermore, the expression and regulation of the *Wnts* and their antagonists could explain the formation of somites. To date, the expression of at least five *Sfrps* have been characterised in either mouse or chick. *Sfrp1*, *Sfrp2* and *Sfrp3* are particularly interesting as they all were expressed in the segmental plate (Hoang et al. 1998; Lee et al. 2000; Terry et al. 2000). Upon somite formation *Sfrp1* and *Sfrp3* expression is down-regulated while the level of *Sfrp2* does not change. Later on *Sfrp2* becomes up-regulated in the sclerotome through the action of Shh (Lee et al. 2000). Importantly, the initiation of *Sfrp2* in the segmental plate is not mediated by Shh (Lee et al. 2000). We suggest that at the level of the segmental plate any epithelialising action of the Wnts is antagonised by the expression of *Sfrp1*, *Sfrp2* and *Sfrp3*, thus maintaining segmental plate in a mesenchymal state. The balance shifts towards epithelialisation in the region of somite formation due to the down-regulation of Wnt antagonist expression (e.g. *Sfrp1* and *Sfrp3*) as well as an increase in *Wnt* expression (Hoang et al. 1998). Subsequently, an up-regulation of *Sfrp2* in the ventral paraxial mesoderm through Shh (Lee et al. 2000) enables the epithelio-mesenchymal transition of the ventral part of the epithelial somite leading to the formation of the sclerotome. In the dorsal part of the somite the epithelial structure (the dermomyotome) is maintained under the influence of Wnt6 and other molecules (e.g. Wnts emanating from the dorsal neural tube). Further experiments have to be carried out to prove this hypothesis.

This works demonstrate that Wnt6 is the epithelialisation factor of the paraxial mesoderm and necessary for proper somite formation.

4.2 The Regulation of *cNkd1* Expression

Paraxial mesoderm undergoes a mesenchymal-to-epithelial transition to form somites. The cells constituting the somites then undergo many cellular changes, however, before they differentiate. At present we have a relatively poor comprehension of how these cellular changes are brought about during the development of the somites.

The Wnt signalling pathway controls numerous aspects of cellular behaviour during both embryonic and adult life, and its components are thus ideal candidates as regulators of somitic development. While it is readily accepted that Wnt signalling leads to the activation of gene transcription (canonical pathway), one often overlooks the ability of these proteins to directly influence cell behaviour through non-canonical mechanisms. It is beyond question that Wnts can control cell behaviour (movement) by regulating the activities of molecules already present within cells. This has been demonstrated in studies of vertebrate gastrulation, where the convergence and extension of an embryo is regulated by Wnts acting in a non-canonical manner (Heisenberg et al. 2000).

We postulated that both canonical and non-canonical pathways could be deployed during somite development. There is considerable evidence that the canonical pathway is used during somite development, particularly in promoting cell division (Galli et al. 2004). In contrast, very few reports have implicated the possibility that Wnt signalling may control the movement of cells during somite development. This is undoubtedly because we have only a sketchy understanding of how the PCP pathway actually works. Recent work, however, has identified the intracellular molecule Nkd as a Wnt-inducible protein that acts to direct canonical signalling towards the PCP pathway (Zeng et al. 2000; Yan et al. 2001).

We have described the expression of chick *Nkd1* showing that it is not only expressed during somite maturation, but its regulation can be modulated by Wnt signalling. *cNkd1* is expressed along almost the entire length of the segmental plate as well as in the somites. Expression in the somites is initially found both in the dorsal and ventral aspect. Expression then becomes localised to the dorsal medial aspect before becoming confined to the dorsal medial lip and the extending myotome.

Our tissue manipulation studies have shown that signals originating from the neural tube are required to maintain the expression of *cNkd1* as the segmental plate undergoes somite formation. Furthermore, we have shown that Wnt1 and Wnt3a are sufficient to maintain *cNkd1* expression in the somites even after their isolation from the neural tube. This raises the issue of identifying the actual member of the Wnt family that initiates *cNkd1* expression in the segmental plate and then maintains its expression in the somites. In order for a particular Wnt to be deemed a good candidate as an initiator of *cNkd1* expression, its own expression profile must correspond to that of the gene it is supposedly initiating. Although two Wnt genes have been previously shown to be expressed in the segmental plate (*Wnt3a* and *Wnt5b*), their expression profiles are not as comprehensive as that of *cNkd1* (Linker et al. 2003). More specifically, *Wnt3a* is only expressed in the posterior half of the segmental plate and Wnt5b is not uniformly expressed in its anterior aspect. Thus, at this stage we can only speculate that these proteins could initiate *cNkd1* expression in the segmental plate. We are on firmer ground when aiming to resolve the question of the identity of the Wnts that maintain the expression of *cNkd1* in the somites. We have shown that the three Wnts expressed in the dorsal neural tube (Wnt1, Wnt3a and Wnt4) are all capable of up-regulating the expression of *cNkd1*. Furthermore, Wnt1 and Wnt3a are sufficient to maintain the expression of *cNkd1* in the paraxial mesoderm after its neural tube isolation.

The expression profile of *cNkd1* during somite development suggests that it could act to regulate cell morphogenesis and movement, since it is expressed in regions of cell ingression during the formation of the myotome. Without further evidence this notion remains merely speculation. Interestingly, Marcelle and co-workers (1997) have proposed that Wnt11 acts to regulate cell movement at the dorsal medial lip during myotome formation (Marcelle et al. 1997). This raises the possibility that Wnt11 could control the expression of *cNkd1*, as they are both thought to regulate the same process. We do not favour this hypothesis since, first, *cNkd1* is expressed in the somites before *Wnt11* and, second, we have shown that Wnt11 actually inhibits the expression of *cNkd1*. We have to explain, however, how neural tube Wnts induce the expression of both *cNkd1* and a protein that will down-regulate *cNkd1*, namely Wnt11. We propose that this can be explained in terms of the temporal expression patterns of these genes. We suggest that during early development dorsal neural tube Wnts positively regulate the expression *cNkd1* in the somites. At later stages the somites become competent to express *Wnt11* under the influence of Wnts from the neural tube. We suggest that at the later stages, Wnt11 may act locally to regulate *cNkd1* expression but is not present at sufficient levels to completely inhibit the expression of *cNkd1*.

If indeed *cNkd1* plays a role in the development of the myotome by regulating cell movement, and if its expression is regulated solely by Wnts originating from the neural tube, this raises two interesting issues concerning axial skeletal muscle. First, note that the expression of *cNkd1* seems to be confined to the epaxial domain, whereas muscle development occurs at the hypaxial domain also. This would suggest that unique signalling pathways are used to regulate the formation of epaxial and hypaxial musculature. Second, our results show that Wnt proteins originating in the neural tube are able to maintain the expression of *cNkd1* in both medial and lateral aspects of the developing myotome. This would imply that the Wnts are capable of acting as long-range signalling molecules.

4.3
Wnts and Neural Crest Induction

A New Role for the Non-canonical Pathway in Neural Crest Induction

Wnt proteins have been implicated as the inducers of neural crest in vertebrates. Wnt6 is the only Wnt known to be expressed in the ectoderm at the precise time of neural crest induction in the chick embryo. It has not been demonstrated, however, that a vertebrate Wnt can induce crest in the amniote embryo in vivo. We have shown for the first time, both in vivo and in vitro, that Wnt6 can fulfil this role in the chick embryo. Importantly, we demonstrated that other Wnts, both canonical and non-canonical, do not perform this role and thus this is not a general property of Wnt molecules. Critically, by specifically inhibiting Wnt6 with siRNA, we have demonstrated that Wnt6 is not only able to induce neural crest but also that it is required for induction in the chick embryo.

Intriguingly, the absence of nuclear activated β-catenin in Wnt6-induced neural crest cells implies that Wnt6 is not acting via the canonical pathway. This hypothesis is further supported through several lines of experimental evidence that we have presented here: (1) expression of an inhibitory Dsh construct that specifically blocks the canonical signalling pathway but leaves the non-canonical pathway active [Δ-DIX-Dvl (Krylova et al. 2000; Rosso et al. 2005)] resulted in increased neural crest cell production. (2) Key components of the non-canonical pathway—RhoA (Kim and Han 2005) and activated Jun kinase (Li et al. 1999)—are ectopically expressed in the dorsal neural tube in Wnt6-implanted embryos. These experiments all indicate that Wnt6 acts non-canonically. This is an unexpected finding given previous studies in anamniotes, such as *Xenopus* and zebrafish, have indicated that while neural crest migration is controlled by non-canonical signalling (De Calisto et al. 2005) the induction of the neural crest is controlled by canonical Wnt signalling (Saint-Jeannet et al. 1997; La Bonne and Bronner-Fraser 1998; Lewis et al. 2004; Abu-Elmagd et al. 2006). These anamniote studies were conducted at earlier stages in development than ours, partly reflecting the earlier production of neural crest in these species, and they mainly analysed anterior trunk crest populations. While it seems unlikely that the requirement for canonical signalling is radically different in more posterior trunk populations of anamniotes, this has yet to be determined.

The apparent discrepancy in the pathways utilised to induce neural crest in amniotes and anamniotes may reflect changes in the specific Wnts and Wnt signalling pathways employed to induce and segregate crest during the evolution of the amniote lineage. In support of our findings, key components of the non-canonical signalling pathway are found to be transiently expressed in the amniote dorsal neural tube around the time of neural crest induction, including RhoA (Liu and Jessell 1998) and Flamingo/Celsr (Formstone and Mason 2005). Equally, in amniote embryos, canonical Wnts (Wnt1 and Wnt3a) are expressed too late to be inducers of neural crest (Dickinson et al. 1995). The presence of nuclear β-catenin, however, has been reported in the open neural plate of the stage 10 chick embryo (Garcia-Castro et al. 2002) which would suggest canonical signalling occurs at the time of crest induction. In our study, we were only able to localise nuclear β-catenin in more cranial regions and not in the open or fusing neural plate between stages 9–12. A possible explanation for this discrepancy is that different antibodies were used in the two analyses. Garcia-Castro et al. (2002) employed 5H10 antibody, which detects pan β-catenin labelling, in both the cytoplasm and the nucleus. In contrast, we have used an antibody which detects only dephosphorylated β-catenin, which is the activated form of the molecule and therefore more likely to be involved in signalling. We find many fewer nuclei are labelled using this antibody and none in the caudal neural tube or plate.

In the same paper it was shown that *Drosophila* Wg can induce chick neural crest in vitro (Garcia-Castro et al. 2002), which would appear to contradict our results. The full spectrum of Wg activity in the amniote, however, is unknown and Wg may be able to activate the amniote non-canonical signalling pathway. Crucially, experiments using Wg to induce neural crest did not examine downstream components to definitively prove which pathway was active.

The exact mechanism by which Wnt6 induces crest from the neuroepithelium remains obscure. We find little difference in cell proliferation rates in the neural tube of Wnt6- and Wnt1-implanted embryos compared to controls (not shown); Wnt6 therefore does not simply increase cell number. The main described function of the non-canonical Wnt signalling pathway is to establish cell polarity, and it has also been implicated in controlling the orientation of cell division during zebrafish gastrulation (Gong et al. 2004). Changes in the plane of cell polarity during cell division are involved in inducing multiple cell types from a single neuroepithelia, such as the retina, and this may lead to cell fate specification (Cayouette and Raff 2003). Neural tube cells predominantly orient their mitotic spindles along the A–P axis; however, delaminating neural crest cells orientate their cell division at 90° to this (Le Douarin and Kalcheim 1999), implying that neural crest cell induction involves changes in cell polarity. It is therefore possible that Wnt6 and the non-canonical pathway change the polarity of cell division in dorsal neural tube cells during crest induction. Induction of neural crest through an alteration in cell polarity may explain an apparent discrepancy between our results and that of Burstyn-Cohen et al. (2004), where inhibition of the non-canonical signalling pathway did not reduce proliferation at the dorsal edge of the neural tube, prompting the authors to suggest that the non-canonical pathway did not have a role in neural crest induction.

The Canonical Pathway Inhibits Neural Crest Induction

A final unexpected finding of our work was that Wnt1 and activation of the canonical pathway inhibited neural crest formation and that removal of non-canonical signalling using a Δ-DEP-Dvl construct, leaving only the canonical active, also resulted in a reduction in neural crest formation. A similar DEP domain deletion construct in *Xenopus* Dsh (dd2; Sokol 1996) has been proposed to inhibit signalling through both the canonical and non-canonical pathways (see De Calisto et al. 2005). While there is no evidence that the constructs we have employed also do this, the resulting phenotype would be identical to the one that we obtained—a significant inhibition in neural crest production, as the neural tube would lack the inducing signal. However, we have also inhibited the function of the canonical pathway, with either dn-TCF-4, Nkd or Δ-DIX-Dvl, all resulting in an increase in neural crest production. While these constructs affect different parts of the signalling pathway, they all indicate that canonical signalling inhibits neural crest production in the chick embryo.

Whilst previous experiments in *Xenopus* (anamniote) embryos indicate that over-expressing *Wnt1* resulted in neural crest formation (Saint-Jeannet et al. 1997), our results show that this is not the case in the chick. It has previously been shown, however, that *dn-Wnt1*-expressing cells inhibit *Slug* expression when implanted in the closing neural plate of the chick embryo (Garcia-Castro et al. 2002). This construct has broad Wnt inhibiting ability (Hoppler et al. 1996), so it is possible that this construct may also be able to inhibit Wnt6 activity, which would result in loss of Slug expression. The activity of this inhibitor requires further investigation.

The activity of Wnt1 and canonical signalling during neural crest induction has not been studied in great detail in other amniotes; however, specific over-expression of β-catenin in the neural crest in mouse embryos resulted in an overall decrease in the amount of neural crest produced, with the remaining neural crest mainly segregating into the sensory neuron lineage (Lee et al. 2004). This suggests that in mice, similar to chick, canonical Wnt signalling can inhibit neural crest induction, implying that this is an amniote characteristic. *Wnt1* is expressed too late to induce neural crest in amniote embryos, being expressed a few hours after the appearance of neural crest markers (Dickinson et al. 1995; Garcia-Castro et al. 2002). Moreover, neural crest is induced in the absence of either Wnt1 or Wnt3a (Dickinson et al. 1995; Ikeya et al. 1997), but later expansion of the neural crest population is affected. Equally, mouse knockouts of β-catenin (Brault et al. 2001; Hari et al. 2002) do not have altered neural crest induction but have altered differentiation and specification of neural crest derived lineages, suggesting that Wnt1 and the canonical pathway are not responsible for neural crest induction. A comprehensive analysis of the role of Wnt1 in chick neural crest delamination used similar constructs to those employed in this study but examined later stages and more anterior crest populations (rostral segmental plate and somite level) to temporally and spatially separate induction and delamination of neural crest (Burstyn-Cohen et al. 2004). This study demonstrated a significant role for Wnt1 and the canonical pathway in the delamination process and also showed that canonical signalling did not affect expression of *FoxD3, Sox9* or *Slug* at these later stages. This suggests a very dynamic role for Wnt1 in chick neural crest development. When neural crest induction commences, it is not immediately expressed, but as it switches on in the caudal neural tube it initially acts as an inhibitor of neural crest production before rapidly changing its role to promote delamination at more anterior levels of the neural tube. It is also interesting to note in *Xenopus* (an anamniote), neural crest delamination is controlled by the non-canonical signalling pathway (De Calisto et al. 2005), suggesting that the downstream signalling pathway controlling delamination in anamniotes and amniotes has also swapped between the canonical and non-canonical pathways.

One interesting finding from our studies was that it appears that Wnt1 and Wnt6 both act in concert during neural crest production to mutually repress each other. Inhibition of the canonical pathway with the Δ-DIX-Dvl construct resulted in a phenotype identical to over-expression of non-canonical Wnt6, namely induction of crest. Conversely, inhibition of the non-canonical pathway, with both inhibiting Dsh construct Δ-DEP-Dvl and Wnt6 siRNA, resulted in a phenotype identical to over-expression of the canonical pathway, namely inhibition of neural crest production. This mutual repression may act as a system to "balance" neural crest production. Mutual repression by a canonical and a non-canonical Wnt is an emerging theme; for example, during axis formation, where non-canonical *Wnt5* is expressed ventrally and canonical *Wnt8* dorsally, removal of *Wnt5* increases canonical signalling and dorsalises the embryo (Westfall et al. 2003), although the exact mechanism by which this repression occurs is unclear. Thus, mutual Wnt repression may be a common mechanism employed throughout the embryo to regulate cell induction, cell differentiation and cell migration.

Acknowledgements

We have to thank Linda Jacobi, Elaine Shervill and Sabine Mühlsimer for excellent technical assistance. These studies were supported by a grant from the Deutsche Forschungsgemeinschaft.

References

Abu-Elmagd M, Garcia-Morales C, Wheeler GN (2006) Frizzled 7 mediates canonical Wnt signalling in neural crest induction. Dev Biol 298:285–298

Artavanis-Tsakonas S, Rand MD, Lake RJ (1999) Notch signalling: cell fate control and signal integration in development. Science 284:770–776

Aulehla A, Johnson RL (1999) Dynamic expression of lunatic fringe suggests a link between notch signalling and an autonomous cellular oscillator driving somite segmentation. Dev Biol 201:49–61

Aulehla A, Wehrle C, Brand-Saberi B, Kemler R, Gossler A, Kanzler B, Herrmann BG (2003) Wnt3a plays a major role in the segmentation clock controlling somitogenesis. Dev Cell 4:395–406

Axelrod J, Miller J, Shulman J, Moon R, Perrimon N (1998) Differential recruitment of Dishevelled provides signalling specificity in the planar cell polarity and Wingless signalling pathways. Genes Dev 12:2610–2622

Bafico A, Liu g, Yaniv A, Gazit A, Aaronson SA (2001) Novel mechanism of Wnt signalling inhibition by Dickkopf-1 interaction with LRP6/Arrow. Nat Cell Biol 3:683–686

Baker CV, Bronner-Fraser M (1997) The origins of the neural crest. Part I: embryonic induction. Mech Dev 69:3–11

Baker JC, Beddington RS, Harland RM (1999) Wnt signalling in *Xenopus* embryos inhibits BMP4 expression and activates neural development. Genes Dev 13:3149–3159

Balling R, Helwig U, Nadeau J, Neubuser A, Schmahl W, Imai K (1996) Pax genes and skeletal development. Ann N Y Acad Sci 785:27–33

Barnes GL, Alexander PG, Hsu CW, Mariani BD, Tuan RS (1997) Cloning and characterization of chicken Paraxis: a regulator of paraxial mesoderm development and somite formation. Dev Biol 189:95–111

Barrantes IB, Elia AJ, Wunsch K, Hrabe de Angelis MH, Mak TW, Rossant J, Conlon RA, Gossler A, de la Pompa JL (1999) Interaction between Notch signalling and Lunatic fringe during somite boundary formation in the mouse. Curr Biol 9:470–480

Bastidas F, De Callisto J, Mayor R (2004) Identification of neural crest competence territory: role of Wnt signalling. Dev Dyn 229:109–117

Bessho Y, Kageyama R (2003) Oscillations, clocks and segmentation. Curr Opin Genet Dev 13:379–384

Bhanot P, Brink M, Samos CH, Hsieh JC, Wang Y, Macke JP, Andrew D, Nathans J, Nusse R (1996) A new member of the frizzled family from *Drosophila* functions as a Wingless receptor. Nature 382:225–230

Bienz M (1998) TCF: transcriptional activator or repressor? Curr Opin Cell Biol 10:366–372

Bienz M, Hamada F (2004) Adenomatous polyposis coli proteins and cell adhesion. Curr Opin Cell Biol 16:528–535

Bonstein L, Elias S, Frank D (1998) Paraxial-fated mesoderm is required for neural crest induction in *Xenopus* embryos. Dev Biol 193:156–168

Borello U, Buffa V, Sonnino C, Melchionna R, Vivarelli E, Cossu G (1999) Differential expression of the Wnt putative receptors Frizzled during mouse somitogenesis. Mech Dev 89:173–177

Borello U, Berarducci B, Murphy P, Bajard L, Buffa V, Piccolo S, Buckingham M, Cossu G (2006) The Wnt/β-catenin pathway regulates Gli-mediated Myf5 expression during somitogenesis. Development 133:3723–3732

Borycki A, Brown AM, Emerson CP Jr (2000) Shh and Wnt signalling pathways converge to control Gli gene activation in avian somites. Development 127:2075–2087

Borycki AG, Mendham L, Emerson CP Jr (1998) Control of somite patterning by sonic hedgehog and its downstream signal response genes. Development 125:777–790

Boutros M, Paricio N, Strutt DI, Mlodzik M (1998) Dishevelled activates JNK and discriminates between JNK pathways in planar cell polarity and wingless signalling. Cell 94:109–118

Brand-Saberi B, Ebensperger C, Wilting J, Balling R, Christ B (1993) The ventralizing effect of the notochord on somite differentiation in chick embryos. Anat Embryol (Berl) 188:239–245

Brand-Saberi B, Wilting J, Ebensperger C, Christ B (1996) The formation of somite compartments in the avian embryo. Int J Dev Biol 40:411–420

Brault V, Moore R, Kutsch S, Ishibashi M, Rowith DH, McMahon AP, Sommer L, Boussadia O, Kemler R (2001) Inactivation of the β-catenin gene by Wnt1-Cre-mediated deletion results in dramatic brain malformation and failure of craniofacial development. Development 128:1253–1264

Bronner-Fraser M (1986) Analysis of the early stages of trunk neural crest migration in avian embryos using monoclonal antibody HNK-1. Dev Biol 115:44–55

Burgess R, Cserjesi P, Ligon KL, Olson EN (1995) Paraxis: a basic helix-loop-helix protein expressed in paraxial mesoderm and developing somites. Dev Biol 168:296–306

Burgess R, Rawls A, Brown D, Bradley A, Olson EN (1996) Requirement of the paraxis gene for somite formation and musculoskeletal patterning. Nature 384:570–573

Burstyn-Cohen T, Stanleigh J, Sela-Donenfeld D, Kalcheim C (2004) Canonical Wnt activity regulates trunk neural crest delamination linking BMP/Noggin signalling with G1/S transition. Development 131:5327–5339

Buttitta L, Tanaka T, Chen A, Ko M, Fan C (2003) Microarray analysis of somitogenesis reveals novel targets of different WNT signalling pathways in the somatic mesoderm. Dev Biol 258:91–104

Carl TF, Dufton C, Hanken J, Klymkowsky MW (1999) Inhibition of neural crest migration in *Xenopus* using antisense Slug RNA. Dev Biol 213:101–115

Cauthen CA, Berdougo E, Sandler J, Burrus LW (2001) Comparative analysis of the expression patterns of Wnts and Frizzled during early myogenesis in chick embryos. Mech Dev 104:133–138

Cavallo RA, Cox RT, Moline M, Roose J, Polevay G, Clevers H, Peifer M, Bejsovec A (1998) *Drosophila* Tcf and Groucho interact to repress Wingless signalling activity. Nature 395:604–608

Cayouette M, Raff M (2003) The orientation of cell division influences cell-fate choice in the developing mammalian retina. Development 130:2329–2339

Chang C, Hemmati-Brivanlou A (1998) Neural crest induction by Xwnt7b in *Xenopus*. Dev Biol 194:129–134

Chen G, Fernandez J, Mische S, Courey AJ (1999) A functional interaction between the histone deacetylase Rpd3 and the corepressor groucho in *Drosophila* development. Genes Dev 13:2218–2230

Cheung M, Briscoe J (2003) Neural crest development is regulated by the transcription factor sox 9. Development 130:5681–5693

Christ B, Ordahl CP (1995) Early stages of chick somite development. Anat Embryol (Berl) 191:381–396
Christ B, Jacob HJ, Jacob M (1974) Origin of wing musculature. Experimental studies on quail and chick embryos. Experientia 30:1446–1449
Christ B, Jacob HJ, Jacob M (1977) Experimental analysis of the origin of the wing musculature in avian embryos. Anat Embryol (Berl) 150:171–186
Christ B, Schmidt C, Huang R, Wilting J, Brand-Saberi B (1998) Segmentation of the vertebrate body. Anat Embryol (Berl) 197:1–8
Christ B, Huang R, Scaal M (2004) Formation and differentiation of the avian sclerotome. Anat Embryol (Berl) 208:333–350
Coles E, Christiansen J, Economou A, Bronner-Fraser M (2004) A vertebrate crossveinless 2 homologue modulates BMP activity and neural crest migration. Development 131:5309–5317
Coles EG, Gammill LS, Miner JH, Bronner-Fraser M (2006) Abnormalities in neural crest migration in laminin alpha5 mutant mice. Dev Biol 289:218–228
Cong F, Schweizer L, Varmus H (2004) Casein Kinase Iε modulates the signalling specificities of Dishevelled. Mol Cell Biol 24:2000–2011
Conlon RA, Reaume AG, Rossant J (1995) Notch 1 is required for the coordinate segmentation of somites. Development 121:1533–1545
Cooke J, Zeeman EC (1976) A clock and wavefront model for control of the number of repeated structures during animal morphogenesis. J Theor Biol 58:455–476
Culi J, Mann RS (2003) Boca, an endoplasmic reticulum protein required for wingless signalling and trafficking of LDL receptor family members in *Drosophila*. Cell 112:343–354
Dale TC (1998) Signal transduction by the Wnt family of ligands. Biochem J 329:209–223
Daniels DL, Weis WI (2005) β-Catenin directly displaces Groucho/TLE repressors from Tcf/Lef in Wnt-mediated transcription activation. Nat Struct Mol Biol 12:364–371
Das R, Van Hateren N, Howell G, Farrell E, Bangs F, Porteous V, Manning E, McGrew M, Ohyama K, Saccor M, et al (2006) A robust system for RNA interference in the chicken using a modified microRNA operon. Dev Biol 294:554–563
De Calisto J, Araya C, Marchant L, Riaz CF, Mayor R (2005) Essential role of non-canonical Wnt signalling in neural crest migration. Development 132:2587–2597
Deardorff MA, Tan C, Saint-Jeannet JP, Klein PS (2001) A role for frizzled 3 in neural crest development. Development 128:3655–3663
Del Barrio MG, Nieto MA (2002) Overexpression of Snail family members highlights their ability to promote chick neural crest formation. Development 129:1583–1593
Delfini MC, Dubrulle J, Malapert P, Chal J, Pourquie O (2005) Control of the segmentation process by graded MAPK/ERK activation in the chick embryo. Proc Natl Acad Sci USA 102:11343–11348
Denetclaw WF, Ordahl CP (2000) The growth of the dermomyotome and formation of early myotome lineages in thoracolumbar somites of chicken embryos. Development 127:893–905
Denetclaw WF jr, Christ B, Ordahl CP (1997) Location and growth of epaxial myotome precursor cells. Development 124:1601–1610
Dickinson ME, Selleck MA, McMahon AP, Bronner-Fraser M (1995) Dorsalization of the neural tube by the non-neural ectoderm. Development 121:2099–2106
Dietrich S, Schubert FR, Lumsden A (1997) Control of dorsoventral pattern in the chick paraxial mesoderm. Development 124:3895–3908
Dietrich S, Abou-Rebyeh F, Brohmann H, Bladt F, Sonnenberg-Rietmacher E, Yamaai T, Lumsden A, Brand-Saberi B, Birchmeier C (1999) The role of SF/HGF and c-Met in the development of skeletal muscle. Development 126:1621–1629

Du SJ, Purcell SM, Christian JL, McGrew LL, Moon RT (1995) Identification of distinct classes and functional domains of Wnts through expression of wild-type and chimeric proteins in *Xenopus* embryos. Mol Cell Biol 15:2625–2634

Dubrulle J, McGrew MJ, Pourquie O (2001) FGF signalling controls somite boundary position and regulates segmentation clock control of spatiotemporal Hox gene activation. Cell 106:219–232

Eastman Q, Grosschedl R (1999) Regulation of LEF-1/TCF transcription factors by Wnt and other signals. Curr Opin Cell Biol 11:233–240

Ebensperger C, Wilting J, Brand-Saberi B, Mizutani Y, Christ B, Balling R, Koseki H (1995) Pax1, a regulator of sclerotome development is induced by notochord and floor plate signals in avian embryos. Anat Embryol (Berl) 191:297–310

Evans DJ (2003) Contribution of somitic cells to the avian ribs. Dev Biol. Anat Embryol (Berl) 256:114–126

Evrard YA, Lun Y, Aulehla A, Gan L, Johnson RL (1998) Lunatic fringe is an essential mediator of somite segmentation and patterning. Nature 394:377–381

Fan CM, Tessier-Lavigne M (1994) Patterning of mammalian somites by surface ectoderm and notochord: evidence for sclerotome induction by a hedgehog homolog. Cell 79:1175–1186

Fan CM, Lee CS, Tessier-Lavigne M (1997) A role for WNT proteins in induction of dermomyotome. Dev Biol 191:160–165

Finch PW, He X, Kelley MJ, Uhren A, Schaudies RP, Popescu NC, Rudikoff S, Aaronson SA, Varmus HE, Rubin JS (1997) Purification and molecular cloning of a secreted Frizzled related antagonist of Wnt action. Proc Natl Acad Sci USA 94:6770–6775

Formstone C, Mason I (2005) Expression of Celsr/flamingo homologue, c-fmi1, in the early avian embryo indicates a conserved role in neural tube closure and additional roles in asymmetry and somitogenesis. Dev Dyn 232:408–413

Forsberg H, Crozet F, Brown NA (1998) Waves of mouse Lunatic fringe expression, in four-hour cycles at two-hour intervals, precede somite boundary formation. Curr Biol 8:1027–1030

Freitas C, Rodrigues S, Saude L, Palmeirim I (2005) Running after the clock. Int J Dev Biol 49:317–324

Galli LM, Willert K, Nusse R, Yablonka-Reuveni Z, Nohno T, Denetclaw W, Burrus LW (2004) A proliferative role for Wnt-3a in chick somites. Dev Biol 269:489–504

Garcia-Castro MI, Marcelle C, Bronner-Fraser M (2002) Ectodermal Wnt function as a neural crest inducer. Science 297:848–851

Geetha-Loganathan P, Nimmagadda S, Proels F, Patel K, Scaal M, Huang R, Christ B (2005) Ectodermal Wnt6 promotes Myf5-dependent avian limb myogenesis. Dev Biol 288:221–233

Geetha-Loganathan P, Nimmagadda S, Huang R, Christ B, Scaal M (2006) Regulation of ectodermal Wnt6 expression by the neural tube is transduced by dermomyotomal Wnt11: a mechanism of dermomyotomal lip sustainment. Development 133:2897–2904

Glinka A, Wu W, Delius H, Monaghan AP, Blumenstock C, Niehrs C (1998) Dickkopf-1 is a member of a new family of secreted proteins and functions in head induction. Nature 391:357–362

Gong Y, Mo C, Fraser SE (2004) Planar cell polarity signalling controls cell division orientation during zebrafish gastrulation. Nature 430:689–693

Gregorieff A, Clevers H (2005) Wnt signalling in the intestinal epithelium: from endoderm to cancer. Genes Dev 19:877–890

Gumbiner BM (1995) Signal transduction of β-catenin. Curr Opin Cell Biol 7:634–640

Hamburger V, Hamilton H (1951) A series of normal stages in the development of the chick embryo. J Morphol 88:49–82

Hari L, Brault V, Kleber M, Lee HY, Ille F, Leimeroth R, Paratore C, Suter U, Kemler R, Sommer L (2002) Lineage-specific requirements of β-catenin in neural crest development. J Cell Biol 159:867–880
Hay ED (1995) An overview of epithelio-mesenchymal transformation. Acta Anat (Basel) 154:8–20
He X, Saint-Jeannet JP, Wang Y, Nathans J, Dawid I, Varmus H (1997) A member of the Frizzled protein family mediating axis induction by Wnt-5a. Science 275:1652–1654
Heisenberg CP, Tada M, Rauch GJ, Saude L, Concha ML, Geisler R, Stemple DL, Smith JC, Wilson SW (2000) Silberblick/Wnt11 mediates convergent extension movements during zebrafish gastrulation. Nature 405:76–81
Hoang BH, Thomas JT, Abdul-Karim F, Coreia K, Conlon RA, Luyten F, Ballock RT (1998) Expression pattern of two Frizzled-related genes, Frzb-1 and Sfrp-1, during mouse embryogenesis suggests a role for modulating action of Wnt family members. Dev Dyn 212:364–372
Hoppler S, Brown JD, Moon RT (1996) Expression of a dominant-negative Wnt blocks induction of MyoD in *Xenopus* embryos. Genes Dev 10:2805–2817
Hrabe de Angelis M, McIntyre J 2nd, Gossler A (1997) Maintenance of somite borders in mice requires the Delta homologue Dll1. Nature 386:717–721
Hsieh JC, Kodjabachian L, Rebbert ML, Rattner A, Smallwood PM, Samos C, Nusse R, Dawid I, Nathans J (1999) A new secreted protein that binds to Wnt proteins and inhibits their activities. Nature 398:431–436
Hsieh JC, Lee L, Zhang L, Wefer S, Brown K, et al (2003) Mesd encodes an LRP5/6 chaperone essential for specification of mouse embryonic polarity. Cell 112:355–367
Huang R, Zhi Q, Wilting J, Christ B (1994) The fate of somitocoel cells in avian embryos. Anat Embryol (Berl) 190:243–250
Huang R, Zhi Q, Neubüser A, Müller TS, Brand-Saberi B, Christ B, Wilting J (1996) Function of somite and somitocoele cells in the formation of the vertebral motion segment in avian embryos. Acta Anat (Basel) 155:231–241
Huang R, Zhi Q, Schmidt C, Wilting J, Brand-Saberi B, Christ B (2000) Sclerotomal origin of the ribs. Development 127:527–532
Huang X, Saint-Jeannet JP (2004) Induction of the neural crest and the opportunities of life on the edge. Dev Biol 275:1–11
Huber O, Korn R, McLaughlin J, Ohsugi M, Herrmann BG, Kemler R (1996) Nuclear localization of β-catenin by interaction with transcription factor LEF-1. Mech Dev 59:3–10
Huelsken J, Behrens J (2002) The Wnt signalling pathway. J Cell Sci 115:3977–3978
Ikeya M, Takada S (1998) Wnt signalling from the dorsal neural tube is required for the formation of the medial dermomyotome. Development 125:4969–4976
Ikeya M, Lee SM, Johnson JE, McMahon AP, Takada S (1997) Wnt signalling required for expansion of neural crest and CNS progenitors. Nature 389:966–970
Itaeranta P, Lin Y, Perasaari J, Roel G, Destree O, Vanio S (2002) Wnt6 is expressed in the ureter bud and induces kidney tubule development in vitro. Genesis 32:259–268
Itasaki N, Jones CM, Mercurio S, Rowe A, Domingos PM, et al (2003) Wise a context-dependent activator and inhibitor of Wnt signalling. Development 130:4295–4305
Jacobson AG (1988) Somitomeres: mesodermal segments of vertebrate embryos. Development 104 Suppl:209–220
Jamora C, DasGupta R, Kocieniewski P, Fuchs E (2003) Links between signal transduction, transcription and adhesion in epithelial bud development. Nature 422:317–322
Jen WC, Wettstein D, Turner D, Chitnis A, Kintner C (1997) The notch ligand, X-Delta-2, mediates segmentation of the paraxial mesoderm in *Xenopus* embryos. Development 124:1169–1178

Johnson J, Rhee J, Parsons SM, Brown D, Olson EN, Rawls A (2001) The anterior/posterior polarity of somites is disrupted in paraxis-deficient mice. Dev Biol 229:176–187

Kawakami Y, Raya A, Raya RM, Rodriguez-Esteban C, Belmonte JC (2005) Retinoic acid signalling links left-right asymmetric patterning and bilaterally symmetric somitogenesis in the zebrafish embryo. Nature 435:165–171

Kiefer JC, Hauschka SD (2001) Myf-5 is transiently expressed in nonmuscle mesoderm and exhibits dynamic regional changes within the presegmental mesoderm and somites I-IV. Dev Biol 232:77–90

Kim GH, Han JK (2005) JNK and ROK alpha function in the noncanonical Wnt/Rhoa signalling pathway to regulate *Xenopus* convergent extension movements. Dev Dyn 232:958–968

Kinzler KW, Vogelstein B (1996) Lessons from hereditary colorectal cancer. Cell 87:159–170

Kishida S, Yamamoto H, Hino S, Ikeda S, Kishida M, Kibuchi A (1999) DIX domains of Dvl and axin are necessary for protein interactions and their ability to regulate β-catenin stability. Mol Cell Biol 19:4414–4422

Kispert A, Vainio S, Shen L, Rowitch DH, McMahon AP (1996) Proteoglycans are required for maintenance of Wnt-11 expression in the ureter tips. Development 12:3627–3637

Kispert A, Vainio S, McMahon AP (1998) Wnt-4 is a mesenchymal signal for epithelial transformation of metanephric mesenchyme in the developing kidney. Development 125:4225–4234

Klingensmith J, Yang Y, Axelrod JD, Beier D, Perrimon N, Sussman DJ (1996) Conservation of Dishevelled structure and function between flies and mice: isolation and characterization of Dvl2. Mech Dev 58:15–26

Krupnik VE, Sharp JD, Jiang C, Robinson K, Chickering TW, et al (1999) Functional and structural diversity of the human Dickkopf gene family. Gene 238:301–313

Krylova O, Messenger MJ, Salinas PC (2000) Dishevelled-1 regulates microtubule stability: a new function mediated by glycogen synthase kinase 3beta. J Cell Biol 151:83–94

Kühl M, Sheldahl LC, Park M, Miller JR, Moon RT (2000) The Wnt/Ca2+ pathway: a new vertebrate Wnt signalling pathway takes shape. Trends Genet 16:279–283

Kusakabe M, Nishida E (2004) The polarity-inducing kinase Par-1 controls *Xenopus* gastrulation in cooperation with 14-3-3 and aPKC. EMBO J 23:4190–4201

La Bonne C, Bronner-Fraser M (1998) Neural crest induction in *Xenopus*: evidence for a two-signal model. Development 125:2403–2414

La Bonne C, Bronner-Fraser M (2000) Snail-related transcriptional repressors are required in *Xenopus* for both the induction of the induction of the neural crest and its subsequent migration. Dev Biol 221:195–205

Ladher RK, Church VL, Allen S, Robson L, Abdelfattah A, Brown NA, Hattersley G, Rosen V, Luyten FP, Dale L, Francis-West PH (2000) Cloning and expression of the Wnt antagonists Sfrp2 and Frzb during chick development. Dev Biol 218:183–198

Le Douarin NM (2004) The avian embryo as a model to study the development of the neural crest: a long and still ongoing story. Mech Dev 121:1089–1102

Le Douarin NM, Kalcheim C (1999) The neural crest. Cambridge University Press, New York

Lee CS, Buttita LA, May NR, Kispert A, Fan CM (2000) SHH-N upregulates Sfrp2 to mediate its competitive interaction with Wnt1 and Wnt4 in the somitic mesoderm. Development 127:109–118

Lee HY, Kleber M, Hari L, Brault V, Suter U, Taketo MM, Kemler R, Sommer L (2004) Instructive role of Wnt/β-catenin in sensory fate specification in neural crest stem cells. Science 303:1020–1023

Leimeister C, Bach A, Gessler M (1998) Developmental expression patterns of mouse sFRP genes encoding members of the secreted frizzled related protein family. Mech Dev 75:29–42

Lescher B, Haenig B, Kispert A (1998) sfrp2 is a target of the Wnt-4 signalling pathway in the developing metanephric kidney. Dev Dyn 213:440–451
Lewis JL, Bonner J, Modrell M, Ragland JW, Moon RT, Dorsky RI, Raible DW (2004) Reiterated Wnt signalling during zebrafish neural crest development. Development 131:1299–1308
Li L, Yuan H, Xie W, Mao J, Caruso AM, McMahon AP, Sussmann DJ, Wu D (1999) Dishevelled proteins lead to two signalling pathways. Regulation of LEF-1 and c-Jun N-terminal kinase in mammalian cells. J Biol Chem 274:129–134
Liem KF Jr, Tremml G, Roelink H, Jessel TM (1995) Dorsal differentiation of neural plate cells induced by BMP-mediated signals from epidermal ectoderm. Cell 82:969–979
Lin K, Wang S, Julius MA, Kitajewski J, Moos M Jr, Luyten FP (1997) The cysteine-rich frizzled domain of Frzb-1 is required and sufficient for modulation of Wnt signalling. Proc Natl Acad Sci USA 94:11196–11200
Linker C, Lesbros C, Stark MR, Marcelle C (2003) Intrinsic signals regulate the initial steps of myogenesis in vertebrates. Development 130:4797–4807
Linker C, Lesbros C, Gros J, Burrus LW, Rawls A, Marcelle C (2005) β-catenin-dependent Wnt signalling controls the epithelial organisation of somites through the activation of paraxis. Development 132:3895–3905
Liu C, Li Y, Semenov M, Han C, Baeg GH, Tan Y, Zhang Z, Lin X, He X (2002) Control of beta catenin phosphorylation/degradation by a dual-kinase mechanism. Cell 108:837–847
Liu JP, Jessell TM (1998) A role for rhoB in the delamination of neural crest cells from the dorsal neural tube. Development 125:5055–5067
Mao B, Niehrs C (2003) Kremen 2 modulates Dickkopf 2 activity during Wnt/LRP6 signalling. Gene 302:179–183
Mao B, Wu W, Davidson G, Marhold J, Li M, et al (2002) Kremen proteins are Dickkopf receptors that regulate Wnt/β-catenin signalling. Nature 417:664–667
Marcelle C, Stark MR, Bronner-Fraser M (1997) Coordinate actions of BMPs, Wnts, Shh and noggin mediate patterning of the dorsal somite. Development 124:3955–3963
Marchant L, Linker C, Ruiz P, Guerro N, Mayor R (1998) The inductive properties of mesoderm suggest that the neural crest cells are specified by a BMP gradient. Dev Biol 198:319–329
Matsubayashi H, Sese S, Lee JS, Shirakawa T, Iwatsubo T, Tomita T, Yanagawa S (2004) Biochemical characterization of the *Drosophila* wingless signalling pathway based on RNA interference. Mol Cell Biol 24:2012–2024
Mayor R, Morgan R, Sargent MG (1995) Induction of the prospective neural crest of *Xenopus*. Development 121:767–777
Mayr T, Deutsch U, Kuhl M, Drexler HC, Lottspeich F, Deutzmann R, Wedlich D, Risau W (1997) Fritz: a secreted frizzled-related protein that inhibits Wnt activity. Mech Dev 63:109–125
McGonnell IM, Graham A (2002) Trunk neural crest has skeletogenic potential. Curr Biol 12:767–771
McKendry R, Hsu SC, Harland RM, Grosscheldl R (1997) LEF-1/TCF proteins mediate wnt-inducible transcription from the *Xenopus* nodal-related 3 promoter. Dev Biol 192:420–431
McMahon AP, Gavin BJ, Parr B, Bradley A, McMahon JA (1992) The Wnt family of cell signalling molecules in postimplantation development of the mouse. Ciba Found Symp 165:199–212
Meier S (1979) Development of the chick embryo mesoblast. Formation of the embryonic axis and establishment of the metameric pattern. Dev Biol 73:24–45
Melkonyan HS, Chang WC, Shapiro JP, Mahadevappa M, Fitzpatrick PA, Kiefer MC, Tomei D, Umanksy SR (1997) SARPs: a family of secreted apoptosis-related proteins. Proc Natl Acad Sci USA 94:13636–13641

Mestres P, Hinrichsen K (1976) Zur Histogenese des Somiten beim Hühnchen. J. J Embryol Exp Morphol 36:669–683

Mikels AJ, Nusse R (2006) Purified Wnt5a protein activates or inhibits β-catenin-TCF signalling depending on receptor context. PloS Biol 4:e115

Molenaar M, van de Wetering M, Oosterwegel M, Peterson-Maduro J, Godsave S, Korinek V, Roose J, Destree O, Clevers H (1996) XTcf-3 transcription factor mediates β-catenin-induces axis formation in *Xenopus* embryos. Cell 86:391–399

Monaghan AP, Kioschis P, Wu W, Zuniga A, Bock D, Poustka A, Delius H, Niehrs C (1999) Dickkopf genes are co-ordinately expressed in mesodermal lineages. Mech Dev 7:45–56

Monsoro-Burq AH, Fletcher RB, Harland RM (2003) Neural crest induction by paraxial mesoderm in *Xenopus* embryos requires FGF signals. Development 130:3111–3124

Moon RT, Brown JD, Yang-Snyder JA, Miller JB (1997) Structurally related receptors and antagonists compete for secreted Wnt ligands. Cell 88:725–728

Moon RT, Bowerman B, Boutros M, Perrimon N (2002) The promise ad perils of Wnt signalling through β-catenin. Science 296:1644–1646

Moon RT, Kohn AD, De Ferrari GV, Kaykas A (2004) WNT and β-catenin signalling: diseases and therapies. Nat Rev Genet 5:691–701

Moreno TA, Kintner C (2004) Regulation of segmental patterning by retinoic acid signalling during *Xenopus* somitogenesis. Dev Cell 6:205–218

Moriguchi T, Kawachi K, Kamakura S, Masuyama N, Yamanaka H, Matsumoto K, Kikuchi A, Nishida E (1999) Distinct domains of mouse Dishevelled are responsible for the c-Jun N-terminal kinase/stress-activated protein kinase activation and the axis formation in vertebrates. J Biol Chem 274:30957–30962

Münsterberg AE, Kitajewski J, Bumcrot DA, McMahon AP, Lassar AB (1995) Combinatorial signalling by Sonic hedgehog and Wnt family members induces myogenic bHLH gene expression in the somite. Genes Dev 9:2911–2922

Nakamura H, Funahashi J (2001) Introduction of DNA into chick embryos in ovo electroporation. Methods 24:43–48

Nieto MA, Sargent MG, Wilkinson DG, Cooke J (1994) Control of cell behaviour during vertebrate development by Slug, a zinc finger gene. Science 264:835–839

Nieto MA, Patel K, Wilkinson DG (1996) In situ hybridization analysis of chick embryos in whole mount and tissue sections. Methods Cell Biol 51:219–235

Oishi I, Suzuki H, Onishi N, Takada R, Kani S, Ohkawara B, et al (2003) The receptor tyrosine kinase Ror2 is involved in non-canonical Wnt5a/JNK signalling pathway. Genes Cells 8:645–654

Olivera-Martinez I, Thelu J, Teillet MA, Dhouailly D (2001) Dorsal dermis development depends on a signal from the dorsal neural tube, which can be substituted by Wnt-1. Mech Dev 100:233–244

Ordahl CP, Le Douarin NM (1992) Two myogenic lineages within the developing somite. Development 114:339–353

Ossipova O, Dhawan S, Sokol S, Green JB (2005) Distinct Par-1 proteins function in different branches of Wnt signalling during vertebrate development. Dev Cell 8:829–841

Oyama T, Kanai Y, Ochiai A, Akimoto S, Oda T, Yanagihara K, Nagafuchi A, Tsukita S, Shibamoto S, Ito F, et al (1994) A truncated β-catenin disrupts the interaction between E-cadherin and alpha-catenin: a cause of loss of intercellular adhesiveness in human cancer cell lines. Cancer Res 54:6282–6287

Palmeirim I, Henrique D, Ish-Horowicz D, Pourquie O (1997) Avian hairy gene expression identifies a molecular clock linked to vertebrate segmentation and somitogenesis. Cell 91:639–648

Palmeirim I, Dubrulle J, Henrique D, Ish-Horowitz D, Pourquie O (1998) Uncoupling segmentation and somitogenesis in the chick presomitic mesoderm. Dev Genet 23:77–85
Pandur P, Lasche M, Eisenberg LM, Kühl M (2002a) Wnt-11 activation of a non-canonical Wnt signalling pathway is required for cardiogenesis. Nature 418:636–641
Pandur P, Maurus D, Kühl M (2002b) Increasingly complex: new player enter the Wnt signalling network. Bioessays 24:881–884
Parr BA, McMahon AP (1995) Dorsalizing signal Wnt-7a required for normal polarity of D-V and A-P axes of mouse limb. Nature 374:350–353
Parr BA, Shea MJ, Vassileva G, McMahon AP (1993) Mouse Wnt genes exhibit discrete domains of expression in the early embryonic CNS and limb buds. Development 119:247–261
Parr BA, Avery EJ, Cygan JA, McMahon AP (1998) The classical mouse mutant postaxial hemimelia results from a mutation in the Wnt 7a gene. Dev Biol 292:228–234
Peifer M, Polakis P (2000) Wnt signalling in oncogenesis and embryogenesis—a look outside the nucleus. Science 287:1606–1609
Peters JM, McKay RM, McKay JP, Graff JM (1999) Casein kinase I transduces Wnt signals. Nature 401:345–350
Pfeffer PL, De Robertis EM, Izpisua-Belmonte JC (1997) Crescent, a novel chick gene encoding a Frizzled-like cysteine-rich domain, is expressed in anterior regions during early embryogenesis. Int J Dev Biol 41:449–458
Piccolo S, Agius E, Leyns L, Bhattacharyya S, Grunz H, Bouwmeester T, De Robertis EM (1999) The head inducer Cerberus is a multifunctional antagonist of Nodal, BMP and Wnt signals. Nature 397:707–710
Polakis P (2002) Casein kinase 1: a Wnt'er of disconnect. Curr Biol 12:R499–R501
Pourquie O (2000) Vertebrate segmentation: is cycling the rule? Curr Opin Cell Biol 12:747–751
Radtke F, Clevers H (2005) Self-renewal and cancer of the gut. Two sides of a coin. Science 307:1904–1909
Rattner A, Hsieh JC, Smallwood PM, Gilbert DJ, Copeland NG, Jenkins NA, Nathans J (1997) A family of secreted proteins contains homology to the cysteine-rich ligand binding domain of frizzled receptors. Proc Natl Acad Sci USA 94:2859–2863
Reya T, Clevers H (2005) Wnt signalling in stem cells and cancer. Nature 434:843–850
Ridgeway AG, Petropoulos H, Wilton S, Skerjanc IS (2000) Wnt signalling regulates the function of MyoD and myogenin. J Biol Chem 275:32398–32405
Rodriguez-Niedenführ M, Dathe V, Jacob HJ, Pröls F, Christ B (2003) Spatial and temporal pattern of Wnt6 expression during chick development. Anat Embryol (Berl) 206:447–451
Roelink H, Wagenaar E, Lopes da Silva S, Nusse R (1990) Wnt-3, a gene activated by proviral insertion in mouse mammary tumors, is homologous to int-1/Wnt-1 and is normally expressed in mouse embryos and adult brain. Proc Natl Acad Sci USA 87:4519–4523
Rong PM, Teillet MA, Ziller C, Le Douarin NM (1992) The neural tube/notochord complex is necessary for vertebral but not limb and body wall striated muscle differentiation. Development 115:657–672
Rosso S, Sussman D, Wynshaw-Boris A, Salinas P (2005) Wnt signalling through Dishevelled, Rac and JNK regulates dendritic development. Nat Neurosci 8:34–42
Rousset R, Mack JA, Wharton KA Jr, Axelrod JD, Cadigan KM, Fish MP, Nusse R, Scott MP (2001) Naked cuticle targets Dishevelled to antagonize Wnt signal transduction. Genes Dev 15:658–671
Saint-Jeannet JP, Varmus HE, Dawid IB (1997) Regulation of dorsal fate in the neuraxis by Wnt-1 and Wnt-3a. Proc Natl Acad Sci USA 94:13713–13718

Sakanaka C, Leong P, Xu L, Harrison SD, Williams LT (1999) Casein kinase iepsilon in the wnt pathway: regulation of β-catenin function. Proc Natl Acad Sci USA 96:12548–12552

Salic AN, Kroll KL, Evans LM, Kirschner MW (1997) Sizzled: a secreted Xwnt8 antagonist expressed in the ventral marginal zone of *Xenopus* embryos. Development 124:4739–4748

Savagner P (2001) Leaving the neighbourhood: molecular mechanisms involved during epithelial-mesenchymal transition. Bioessays 23:912–923

Scaal M, Bonafede A, Dathe V, Sachs M, Cann G, Christ B, Brand-Saberi B (1999) SF/HGF is a mediator between limb patterning and muscle development. Development 126:4885–4893

Schmidt C, Christ B, Patel K, Brand-Saberi B (1998) Experimental induction of BMP-4 expression leads to apoptosis in the paraxial mesoderm and lateral plate mesoderm. Dev Biol 202:253–263

Schmidt C, Stoeckelhuber M, McKinnell I, Putz R, Christ B, Patel K (2004) Wnt6 regulates the epithelialisation process of the segmental plate mesoderm leading to somite formation. Dev Biol 271:198–209

Schmidt C, Otto A, Luke G, Valasek P, Otto WR, Patel K (2006) Expression and regulation of Nkd-1, an intracellular component of Wnt signalling pathway in the chick embryo. Anat Embryol (Berl) 211:525–534

Schmidt M, Tanaka M, Münsterberg A (2000) Expression of β-catenin in the developing chick myotome is regulated by myogenic signals. Development 127:4105–4113

Schnell S, Maini PK (2000) Clock and induction model for somitogenesis. Dev Dyn 217:415–420

Schubert FR, Mootoosamy R, Walters E, Graham A, Tumiotto L, Muensterberg A, Lumsden A, Dietrich S (2002) Wnt6 marks sites of epithelial transformation in the chick embryo. Mech Dev 114:143–148

Sela-Donenfeld D, Kalcheim C (1999) Regulation of the onset of neural crest migration by coordinated activity of BMP4 and Noggin in the dorsal neural tube. Development 126:4749–4762

Sela-Donenfeld D, Kalcheim C (2000) Inhibition of Noggin expression in the dorsal neural tube by somitogenesis: a mechanism for coordinating the timing of neural crest emigration. Development 127:4845–4854

Selleck MA, Bronner-Fraser M (1995) Origins of the avian neural crest: the role of neural plate-epidermal interactions. Development 121:525–538

Selleck MA, Bronner-Fraser M (2000) Avian neural crest cell fate decisions: a diffusible signal mediates induction of neural crest by the ectoderm. Int J Dev Neurosci 18:621–627

Semenov MV, Snyder M (1997) Human Dishevelled genes constitute a DHR-containing multigene family. Genomics 42:302–310

Semenov MV, Tamai K, Brott BK, Kuehl M, Sokol S, He X (2001) Head inducer Dickkopf-1 is a ligand for Wnt coreceptor LRP6. Curr Biol 11:951–961

Shirozu M, Tada H, Tashiro K, Nakamura T, Lopez ND, Nazarea M, Hamada T, Sato T, Nakano T, Honjo T (1996) Characterization of novel secreted and membrane proteins isolated by the signal sequence trap method. Genomics 37:273–280

Sirbu IO, Duester G (2006) Retinoic-acid signalling in node ectoderm and posterior neural plate directs left-right patterning of somatic mesoderm. Nat Cell Biol 8:271–277

Sokol SY (1996) Analysis of Dishevelled signalling pathways during *Xenopus* development. Curr Biol 6:1456–1467

Sokol SY (1999) Wnt signalling and dorso-ventral axis specification in vertebrates. Curr Opin Genet Dev 9:405–410

Sosic D, Brand-Saberi B, Schmidt C, Christ B, Olson EN (1997) Regulation of paraxis expression and somite formation by ectoderm- and neural tube-derived signals. Dev Biol 185:229–243

Sun TQ, Lu B, Feng JJ, Reinhard C, Jan YN, Fantl WJ, Williams LT (2001) PAR-1 is a Dishevelled-associated kinase and a positive regulator of Wnt signalling. Nat Cell Biol 3:628–636

Sussman DJ, Klingensmith J, Salinas P, Adams P, Nusse R, Perrimon N (1994) Isolation and characterization of a mouse homolog of the *Drosophila* segment polarity gene Dishevelled. Dev Biol 166:73–86

Tajbakhsh S, Borello U, Vivarelli E, Kelly R, Papkoff J, Duprez D, Buckingham M, Cossu G (1998) Differential activation of Myf5 and MyoD by different Wnts in explants of mouse paraxial mesoderm and the later activation of myogenesis in the absence of Myf5. Development 125:4155–4162

Takada S, Stark KL, Shea MJ, Vassileva G, McMahon JA, McMahon AP (1994) Wnt-3a regulates somite and tailbud formation in the mouse embryo. Genes Dev 8:174–189

Takeda H, Lyle S, Lazar AJ, Zouboulis C, Smyth J, Watt FM (2006) Human sebaceous tumors harbour inactivating mutations in LEF-1. Nat Med 12:395–397

Tamai K, Semenov M, Kato Y, Spokony R, Liu C, Katsuyama Y, Hess F, Saint-Jeannet JP, He X (2000) LDL-receptor-related proteins in Wnt signal transduction. Nature 407:530–535

Terry K, Magan H, Baranski M, Burrus M (2000) Sfrp-1 and sfrp2 are expressed in overlapping and distinct domains during chick development. Mech Dev 97:177–182

Tetsu O, McCormick F (1999) β-Catenin regulates expression of cyclin D1 in colon carcinoma cells. Nature 398:422–426

Thomas KR, Capecchi MR (1990) Targeted disruption of the murine int-1 proto-oncogene resulting in severe abnormalities in midbrain and cerebellar development. Nature 346:847–850

Tree DR, Shulman J, Rousset R, Scott M, Gubb D, Axelrod J (2002) Prickle mediates feedback amplification to generate asymmetric planar cell polarity signalling. Cell 109:371–381

Tsang M, Lijam N, Yang Y, Beier DR, Wynshaw-Boris A, Sussman DJ (1996) Isolation and characterization of mouse Dishevelled-3. Dev Dyn 207:253–262

Vallin J, Thuret R, Giacomello E, Faraldo MM, Thiery JP, Broders F (2001) Cloning and characterization of three *Xenopus* slug promoter reveal direct regulation by Lef/β-catenin signalling. J Biol Chem 32:30350–30358

Veeman MT, Slusarski DC, Kaykas A, Louie SH, Moon RT (2003) Zebrafish prickle, a modulator of noncanonical Wnt/Fz signalling, regulates gastrulation movements. Curr Biol 13:680–685

Vermot J, Pourquie O (2005) Retinoic acid coordinates somitogenesis and left-right patterning in vertebrate embryos. Nature 435:215–220

Vermot J, Gallego Llamas J, Fraulob V, Niederreither K, Chambon P, Dolle P (2005) Retinoic acid controls the bilateral symmetry of somite formation in the mouse embryo. Science 308:563–566

Villanueva S, Glavic A, Ruiz P, Mayor R (2002) Posteriorization by FGF, Wnt and retinoic acid is required for neural crest induction. Dev Biol 241:289–301

Wagner J, Schmidt C, Nikowits W, Christ B (2000) Compartmentalization of the somite and myogenesis in chick embryos are influenced by wnt expression. Dev Biol 228:86–94

Wallin J, Wilting J, Koseki H, Fritsch R, Christ B, Balling R (1994) The role of Pax1 in axial skeleton development. Development 120:1109–1121

Wehrli M, Dougan ST, Caldwell K, O'Keefe L, Schwartz S, et al (2000) Arrow encodes an LDL-receptor-related protein essential for Wingless signalling. Nature 407:527–530

Westfall TA, Brimeyer R, Twedt J, Gladon J, Oberding A, Furutani-Seiki M, Slusarski DC (2003) Wnt-5/pipetail functions in vertebrate axis formation as a negative regulator of Wnt/β-catenin activity. J Cell Biol 162:889–898

Weston CR, Davis RJ (2002) The JNK signal transduction pathway. Curr Opin Genet Dev 12:14–21

Wharton KA (2003) Runnin' with the Dvl: proteins that associate with Dsh/Dvl and their significance to Wnt signal transduction. Dev Biol 253:1–17

Wharton KA, Zimmermann G, Rousset R, Scott MP (2001) Vertebrate proteins related to *Drosophila* Naked cuticle bind Dishevelled and antagonize Wnt signaling. Dev Biol 234:93–106

Wiggan O, Hamel PA (2002) Pax3 regulates morphogenetic cell behaviour in vitro coincident with activation of a PCP/non-canonical Wnt-signalling cascade. J Cell Sci 115:531–541

Wilkinson D, Bailes J, McMahon AP (1987) Expression of the proto-oncogene int-1 is restricted to specific neural cells in the developing mouse embryo. Cell 50:79–88

Willert K, Nusse R (1998) β-catenin: a key mediator of Wnt signalling. Curr Opin Genet Dev 8:95–102

Willert K, Brink M, Wodarz A, Varmus H, Nusse R (1997) Casein kinase 2 associates with and phosphorylates Dishevelled. EMBO J 16:3089–3096

Williams BA, Ordahl CP (1994) Pax3 expression in segmental mesoderm marks early stages in myogenic cell specification. Development 120:785–796

Wilson-Rawls J, Hurt CR, Parsons SM, Rawls A (1999) Differential regulation of epaxial and hypaxial muscle development by paraxis. Development 126:5217–5229

Wodarz A, Nusse R (1998) Mechanisms of Wnt signalling in development. Annu Rev Cell Biol 14:59–88

Yamaguchi TP, Bradley A, McMahon AP, Jones S (1999) A Wnt5a pathway underlies outgrowth of multiple structures in the vertebrate embryo. Development 126:1211–1223

Yan D, Wallingford JB, Sun TQ, Nelson AM, Sakanaka C, Reinhard C, Harland RM, Fantl WJ, Williams LT (2001) Cell autonomous regulation of multiple Dishevelled-dependent pathways by mammalian nkd. Proc Natl Acad Sci USA 98:3802–3807

Yoshikawa S, McKinnon RD, Kokel M, Thomas JB (2003) Wnt-mediated axon guidance via the *Drosophila* Derailed receptor. Nature 422:583–588

Yoshikawa Y, Fujimori T, McMahon AP, Takada S (1997) Evidence that absence of Wnt-3a signalling promotes neuralization instead of paraxial mesoderm development in the mouse. Dev Biol 183:234–242

Zeng W, Wharton KA, Mack JA, Wang K, Gadbaw M, Suyama K, Klein PS, Scott MP (2000) Naked cuticle encodes an inducible antagonist of Wnt signalling. Nature 403:789–795

Printing: Krips bv, Meppel, The Netherlands
Binding: Stürtz, Würzburg, Germany